A Manager's Guide for Better Decision-Making

Manufacturing and Production Engineering Series

Series Editors: Hamid R. Parsaei, Texas A&M University at Qatar, & Waldemar Karwowski, University of Central Florida

This series will provide an outlet for the state-of-the-art topics in manufacturing and production engineering disciplines. This new series will also provide a scientific and practical basis for researchers, practitioners, and students involved in areas within manufacturing and production engineering. Issues envisioned to be addressed in this new series would include, but not limited to, the following: Additive Manufacturing, 3D Visualization, Mass Customization, Material Processes, Cybersecurity, Data Science, Automation and Robotics, Underwater Autonomous Vehicles, Unmanned Autonomous Vehicles, Robotics and Automation, Six Sigma and Total Quality Management, Manufacturing Cost Estimation and Cost Management, Industrial Safety, Programmable Logic Controllers, to name just a few.

Decision Making in Risk Management
Quantifying Intangible Risk Factors in Projects
Christopher Cox

Reconfigurable Manufacturing Enterprises for Industry 4.0
Ibrahim H. Garbie and Hamid Parsaei

Recent Advances in Time Series Forecasting
Dinesh C. S. Bisht and Mangey Ram

For more information on this series, please visit: https://www.routledge.com/Manufacturing-and-Production-Engineering/book-series/CRCMNPRDENG

A Manager's Guide for Better Decision-Making

Easy to Apply Tools and Techniques

Abu S. M. Masud

CRC Press
Taylor & Francis Group
Boca Raton London New York

CRC Press is an imprint of the
Taylor & Francis Group, an **informa** business

First edition published 2022
by CRC Press
6000 Broken Sound Parkway NW, Suite 300, Boca Raton, FL 33487-2742

and by CRC Press
2 Park Square, Milton Park, Abingdon, Oxon, OX14 4RN

Library of Congress Cataloging-in-Publication Data

Names: Masud, Abu Syed M., 1947- author.
Title: A manager's guide for better decision-making : easy to apply tools
and techniques / Abu S.M. Masud.
Description: Boca Raton, FL : CRC Press, 2022. | Series: Manufacturing and
production engineering | Includes bibliographical references and index.
Identifiers: LCCN 2021020703 (print) | LCCN 2021020704 (ebook) | ISBN
9781032000169 (hardback) | ISBN 9781032000145 (paperback) | ISBN
9781003172291 (ebook)
Subjects: LCSH: Decision making. | Problem solving.
Classification: LCC HD30.23 .M3723 2022 (print) | LCC HD30.23 (ebook) |
DDC 658.4/03—dc23
LC record available at https://lccn.loc.gov/2021020703
LC ebook record available at https://lccn.loc.gov/2021020704

ISBN: 978-1-032-00016-9 (hbk)
ISBN: 978-1-032-00014-5 (pbk)
ISBN: 978-1-003-17229-1 (ebk)

DOI: 10.1201/9781003172291

Typeset in Times LT Std
by KnowledgeWorks Global Ltd.

Dedication

This book is dedicated to
My wife Aiman Masud
and
Our daughters Farihah Ali and Maha Masud

Contents

Acknowledgments

Several individuals have helped me during the preparation of this book, and a simple thank you is not enough to acknowledge their contributions. Dr Hamid Parsaei, a friend and a colleague as well as the Series Editor of this book, has been a constant source of encouragement. My friend Dr A. M. M. Jamal is one person I could always call, have long discussions about the book, and get his unvarnished editorial judgments. Three of my family members, our two daughters – Farihah Ali and Maha Masud – and our son-in-law Naushad Ali, have read various parts of the book and have provided editorial comments from their perspectives. Cindy Carelli and Erin Harris, the editorial team at CRC Press, have been very helpful from the inception to the publication of this book.

Over the years, I have used much of the material in the book in my domestic and international seminars and workshops. These seminar and workshop participants from many different types of private and public sector organizations have provided valuable feedback about how I could make the material more applicable to their organizations. I have used their suggestions to curate the book content. If the reader finds the book useful then all the credits go to these seminar and workshop participants. Finally, a sincere thank you to all the readers of this book for your confidence in me.

—**Abu S. M. Masud**

Preface

An important activity of an organization's leaders and managers is making decisions associated with organizational problem-solving. All such decisions involve a conceptualization of the choice situation in the form of a model and some simplification of reality. No matter the simplifications made, the process comprises a situation or scenario where the decision-maker (DM) must choose from the possible courses of action in an environment where the available resources are limited and in the presence of various constraints (legal, corporate, regulatory, contractual, physical, etc.). While decision-making methods can be grouped under intuitive approaches, programmed approaches, and analytic approaches, this book focuses on the analytic approaches. Analytic approaches are based on a structured and systematic process.

Some organizations may have external or internal analysts to assist in problem analysis, data collection, decision alternatives development, and alternatives evaluation. Many medium to small organizations may not have these resources for the DM to rely on. Irrespective of whether and what type of analysis assistance the DM may have available, they may still want to do some quick and dirty analysis of their own to reinforce confidence in the alternative finally chosen for implementation.

A wide-ranging amount of decision support materials is available in books and other publications as well as in different sources on the web. However, much of that material may not be easily understood by, and may even be overwhelming for, most DMs when they simply want to apply them. Therefore, this book focuses on providing the DM with some easy-to-apply tools and techniques to conduct their own quick and dirty analysis to identify a 'good' solution. We have chosen only those tools and techniques that we have found useful and easily understood by DMs from a wide range of public and private sector enterprises who were participants in the domestic and international workshops that we have presented. The tools and techniques explained here are those that are time-tested and are tried and true. We have tried to keep the level of mathematics used as low as possible but, as the techniques are quantitative, some familiarity with, and an understanding of, relatively simple mathematics knowledge will be needed.

Chapter One of the book provides an overview of some basic concepts and definitions that are needed to understand the rest of the book. Also, one public-sector problem and one private-sector problem are used to explain the common elements of any decision problem, which are multiple alternatives, multiple and conflicting criteria for evaluating alternatives, and the need for trade-offs or compromise among outcomes.

In *Chapter Two,* some common tools for process and problem analysis are explained. The tools explained are flowchart, check sheets, histogram, Pareto chart, cause-and-effect diagram, affinity diagram, and tree diagram.

In *Chapter Three,* we have described some approaches to the development of alternatives and evaluation criteria for a decision problem. The process for developing alternatives and evaluation criteria is similar to a creative process. This chapter describes the following easy-to-apply creativity fostering techniques: idea checklist, brainstorming, SWOT analysis, nominal group technique (NGT), and Delphi technique.

In *Chapter Four*, commonly used measurement scales for outcome achievement, techniques for normalization of outcome evaluations, and methods for developing criteria importance weights are described. Criteria outcomes, measured using differing scales, may have to be normalized to properly conduct trade-offs between the criteria outcomes. Criteria weights may be needed if the criteria outcomes must be combined. Different measurement scales and methods for normalization and developing criteria importance weights are explained with the help of numerical examples.

In *Chapter Five*, solution methods for decision problems that have one evaluation criterion are explained. When there is no uncertainty regarding the outcome values, the solution approach is Single Criterion Optimization method. When there is uncertainty about the future outcomes of the decision to be made, available solution methods are the Laplace method, Max-Min method, Min-Max Regret method, Expected Value method, and Decision Tree method. Numerical examples are provided to explain the solution steps of these methods.

In *Chapter Six*, we describe methods involving multiple criteria for evaluating alternate solutions to a decision problem. These methods can be grouped into two categories: those that do not require any preference or trade-off information from the DM and those that require the DM's preference or trade-off information. Methods that do not require preference information from the DM include Min-Max Regret method and Compromise Programming method. Methods that require the DM's preference information are the Effective Value method, TOPSIS method, and Analytic Hierarchy Process (AHP) method. Each of these methods has been explained through the solution of at least one numerical example.

For solving a realistic-size problem, all the tools and techniques described in the book will need the use of a computer-based spreadsheet program. A listing of some useful resource material is provided at the end of the book in case you need additional information. We hope that you will find this book useful for most of your managerial decision-making.

About the Author

Abu S. M. Masud has received a BSc (Mechanical Engineering, Bangladesh University of Engineering Technology), Diploma in Business Administration (Institute of Business Administration, Bangladesh), and MS and PhD (Industrial Engineering, Kansas State University, USA). He is a licensed Professional Engineer (Kansas, USA). He is an elected Fellow of the Institute of Industrial and Systems Engineers as well as of the Industrial Engineering and Operations Management Society International.

He is currently an Emeritus Professor of Industrial, Systems, and Manufacturing Engineering at Wichita State University, USA. He has 37 years of teaching and academic administration experience (as Dean of the Graduate School, Associate Dean of Engineering, and Department Chair) at Wichita State University. He was also the Boeing Professor of Global Engineering, a chaired position, at the same university. From 1978 to 1980, he was an Assistant Professor of Industrial Engineering and a researcher at the New Mexico Solar Energy Institute of New Mexico State University, USA. For two summers he was ASEE-NASA Summer Faculty Fellow at NASA Langley Research Center.

From 1987 to 1988, he was a Researcher at Kuwait Institute for Scientific Research. He has about 4 years of industrial experience in power generation and gas distribution organizations in Bangladesh. He has consulted extensively for aircraft manufacturers (Boeing, Cessna, Learjet), electric utility company (KG&E), many medium and small-sized manufacturing companies in the USA as well as at academic institutions in the USA and the Middle East. He has given seminars and workshops in the USA and internationally on topics ranging from decision-making, student learning assessment, academic program development, and institutional and program accreditation. Since the mid-1990s, he has been a program evaluator for the Engineering Accreditation Commission of ABET, the engineering program accreditor in the USA. For over 20 years, he was a peer evaluator for the Higher Learning Commission, a regional accreditor of higher learning institutions in the USA. He is the author/coauthor of a book, six book chapters, and over 45 journal articles and conference proceedings papers.

He currently consults in the areas of engineering education (accreditation, program development, and student learning assessment) and the application of engineering principles in private-sector and public-sector organizations (decision-making and analysis, forecasting and data analytics, and system improvement). He also offers short courses and seminars in decision analysis, forecasting, data analytics, and multiple-criteria decision-making.

Abu S. M. Masud's ORCID link is https://orcid.org/0000-0002-6844-0307.

1 Introduction

1.1 OVERVIEW

Procedures for decision-making can be grouped into intuitive, programmed, and analytical types. All these procedure types have their place in decision-making. It is not possible to make flat statements concerning their soundness or effectiveness. *Intuitive decision-making* implies instinctive decisions without a conscious process for the development and evaluation of alternatives. The process is usually the decision-maker's own and internal to them. One of the problems with intuitive decisions is that the decision-maker (DM) cannot justify the choice made and, hence, it is often difficult to persuade others about its appropriateness. However, an intuitive decision may not mean a complete absence of analysis. More likely, though, it may simply mean that the synthesis of analytical and other information is intuitive and internal to the DM. *Programmed decisions* are those that are made automatically, based on previous analysis or experience. For certain problem situations, it is possible to develop guidelines or instructions for decision-making based often on past analysis and experience. Usually, these problems are routine, and the DM has high confidence in the robustness of the solution approach prescribed for it. *Analytical approaches* to decision-making involve structured analysis, often with the help of mathematical models of the possible courses of action and their consequences. Analytic approaches provide structure and guidance for systematic thinking for solving difficult problems. Some of these concepts are discussed in more detail in Morris (1977) and Bunn (1984). In this book, we will focus only on analytical approaches.

Common elements in all decision problems are:

- multiple decision alternatives
- multiple and conflicting criteria for judging alternatives
- varying criteria achievement outcomes of alternatives
- need for trade-off or compromise among outcomes
- potential uncertainty surrounding outcomes and available information

With two real-life example problems, and with minimal analysis, we demonstrate that both the decision problems exhibit similar characteristics: the presence of multiple, conflicting criteria for judging the alternatives or courses of action and the need for making compromises or trade-offs regarding the outcomes of alternate courses of action. There may, however, be other decision problems where the potential alternatives may be judged by one main and important criterion, while other less important criteria may be used as constraints to weed out some potential alternatives. In this book, we will explain selected solution methods for both types of problems.

DOI: 10.1201/9781003172291-1

1.2   TWO EXAMPLES OF COMPLEX DECISION PROBLEMS

Consider, as examples of the complexity involved in decision-making, two decision problems, one each in public and private sectors.

1.2.1   A PUBLIC SECTOR EXAMPLE PROBLEM

This problem involves a county commission trying to develop a solid waste disposal policy. Because the current landfill cannot be used after 5 years, the county commissioners will have to develop a long-term solution for solid waste disposal. The problem involves consideration of competing technologies, the potential location of incineration plant/transfer station/landfill sites, impact on homeowners near these sites, cost to users, environmental/pollution impacts, impact on underground water aquifers and river waters, and impact on property tax. The multiple conflicting decision criteria include, among others, minimization of impacts on homeowners, environment, underground water sources, tax, and user cost. The alternatives are defined by the technologies available and their characteristics, as well as the available sites. In terms of available technologies, trash can be hauled directly to a new burial site, or moved to a transfer station where it may be sorted or temporarily stored before final disposal (to be buried, composted, or burned), or it can be burned at the new site (with or without recovery of some of the generated heat). Thus, one can easily see that multiple alternatives (probably 5–10 in number) would emerge as serious candidates for consideration. Much of the information, data, and cause-effect relations used in the analysis will have some degree of uncertainty associated with them. For example, future trash transportation costs cannot be estimated exactly because it would depend on factors such as oil price, labor cost, and degree of competition among the trash haulers. Information about these variables is not known with certainty, nor the relationships between these variables are known completely. The potential sites would have different outcomes or achievements for the evaluation criteria. That is, one site may, for example, score high for impact on underground aquifers and river waters (because it has a minimal effect), but it may have a mediocre impact on homeowners near the site, and score very poorly in impact on property tax (because of the high cost to acquire and prepare the site). Another site may be mediocre for the impact on underground aquifers and river waters and impact on property tax but score high for the impact on homeowners near the site (because of low noise and smell generated). Even if these two sites are equally attractive for all the remaining criteria, a decision concerning the selection of one of the sites would require the DMs (in this case, the county commission) to make formal or informal trade-offs among the achievements in all evaluation criteria. This is because no site is expected to be the 'best' for all evaluation criteria.

1.2.2   A PRIVATE SECTOR EXAMPLE PROBLEM

This problem concerns a general aviation manufacturer trying to decide where to locate a new assembly plant for a new model line. Some of the considerations in this decision are the availability of a sufficient amount of land area, availability of trained or trainable potential employees, quality of road/rail communication with existing parts manufacturing plants, availability of a general aviation airfield adjacent to the plant, tax incentives available from city/county/state where the plant will be located, quality of schools and other quality of life issues for potential employees, and distance from the main part production facility. While many cities and counties may wish to be considered for the assembly plant location, only a few will be serious candidates because most would lack a suitable general aviation airfield (because assembled planes will need to fly out of the plant!). Even the existence of a suitable airfield is not enough. There must be sufficient land available near the airfield for procurement, development, and future extension of the plant. Finally, the search may be limited to only certain geographic regions (for political considerations, desire to remain near main parts plant, and other reasons). After these minimum requirements are met, only a handful of sites will remain as candidates for consideration. Next, these sites will have to be evaluated, possibly with uncertain and incomplete information, for evaluation criteria that might include net total investment necessary, tax abatement and other inducements available, availability of suitable labor pool, business environment, quality of communication link with main parts plant (e.g., distance, highway and rail link availability), nature of quality of life factors (such as availability of housing, schools and other educational facilities, cost of living, recreational facilities, etc.) for employees, and availability of utilities (gas, electricity, trash disposal, etc.). It would be a miracle if one location emerges as the best for all evaluation criteria. As a result, the company management, the DMs in this case, will have to trade-off achievements in different criteria to arrive at the best compromise location.

1.3 EXAMPLE PROBLEM 1.1

This problem will be used to illustrate some of the concepts described in this chapter.

Consider a decision problem with five feasible alternatives A_1, A_2, A_3, A_4, and A_5, and three criteria (all to be maximized, that is the more the better) C_1, C_2, and C_3. The problem data is provided in Table 1.1. Such a table is often referred to as a pay-off table. Note that C_2 was originally a minimization criterion that has been converted to a maximization criterion by multiplying the outcome achievements by -1.

TABLE 1.1
Example Problem 1.1 Data

Criterion → Alternative ↓	C_1	C_2	C_3
A_1	50	−30	25
A_2	75	−50	7
A_3	40	−35	15
A_4	20	−10	18
A_5	15	−60	5

1.4   SOME DEFINITIONS

To provide a common understanding of decision problems and their solution methods, we provide here definitions of some terms and concepts used in this book. Many of the following definitions are from Hwang and Masud (1979) and Masud and Ravindran (2008).

> *Alternatives*: *Alternatives are the possible courses of action (or potential solutions).*

Alternatives are at the heart of decision-making. In many decision situations, alternatives can be prespecified. In such cases, every attempt must be made for the identification of all feasible alternatives. Failure to do so may result in selecting an alternative for implementation which may be inferior to other unexplored ones. In most real-life problems, except in trivial decision situations, the number of alternatives to choose from is multiple.

> *Attributes*: *Attributes are the traits, characteristics, qualities, or performance parameters of the alternatives.*

For example, if the decision situation is one of choosing the best car to purchase, then the attributes could be color, gas mileage, attractiveness, size, etc. Attributes are, from a decision-making point-of-view, the descriptors of the alternatives. For some problems, attributes can also form the evaluation criteria.

> *Objectives*: *Objectives are the directions of improvement or to do better, as perceived by DM.*

For example, considering the same example of choosing the best car, an objective may be to maximize gas mileage. This objective indicates that the DM prefers higher gas mileage; the higher the better.

> *Goals*: *Goals are the specific (or desired) status of attributes or objectives. Goals are targets or thresholds of objective or attribute values that are expected to be attained by the best alternative.*

For example, in choosing the best car, a goal may be to buy a car that achieves an average gas mileage of 20 mpg or more; another example, prefer a '4-door car'.

> *Criteria*: *Criteria are the rules of acceptability or standards of evaluation for the alternatives. That is, criteria encompass attributes, goals, and objectives.*

A criterion may be *True* or *Surrogate*. When a criterion is directly measurable, it is called a *True* criterion. An example of a true criterion is the 'cost of the car', which is directly measured by its dollar price. When a criterion is not directly measurable, it may be substituted by one or more surrogate criteria. A *Surrogate* criterion is used in place of one or more others that are more expressive of the DM's underlying values but are difficult to measure directly. An example may be using 'headroom for back seat passengers' as a surrogate for 'passenger comfort'. 'Passenger Comfort' is more expressive as a criterion but is very difficult to measure. 'Headroom for Back Seat Passengers' is however easier to measure and can be used as one surrogate criterion for representing 'Passenger Comfort'.

All decision problems usually have constraints.

Constraints: *Constraints are the conditions that a solution must satisfy or meet.*

In other words, constraints are restrictions on the potential solutions imposed due to physical or other similar limitations. For example, the floor space available for a warehouse may be a constraint in a location problem. Some constraints may be 'hard', that is, they must be met, and some may be 'soft', that is, they are adjustable within a limited range, if necessary.

1.5 CONCEPT OF A 'BEST' SOLUTION

In this section, various concepts surrounding the notion of the best solution are described. Some of the following definitions are from Hwang and Masud (1979) and Masud and Ravindran (2008).

While the focus of the DM may be to identify the best solution for implementation, it must also be a feasible solution.

Feasible Solution: *A feasible solution (or alternative) is a solution that meets (or satisfies) all applicable constraints.*

In single criterion decision problems, the 'Best' solution is defined in terms of an 'Optimum Solution'.

Optimum Solution: *An optimum solution is a feasible solution for which the criterion value is maximum (if the criterion implies the more the better) or minimum (if the criterion implies the less the better) compared to all other alternatives in the set of feasible alternatives.*

In multiple criteria decision problems, however, since all criteria optimums do not usually point to the same alternative, a conflict exists. The notion of an 'Optimum Solution' does not exist in the context of conflicting, multiple criteria. Decision-making for such a problem is, thus, equivalent to choosing the 'most preferred' or the 'best compromise' solution. Even in problems with single criterion, the search for the optimum solution may be too difficult to achieve in a reasonable time or manner. In such cases, we may have to settle for the best achievable solution.

In the absence of an optimal solution, the concepts of dominated and non-dominated solution may become relevant. Also, concepts of ideal solution and anti-ideal solution may be relevant in some methods. These concepts will be explained using the data in Example Problem 1.1. Note that in our discussions throughout the book we will generally assume that all criteria are benefit-type, which are to be max-imized. Keep in mind that in general, one or more criteria can be cost-type which is to be minimized. Since mathematically 'Minimize criterion C is equal to Maximize criterion (-C)', the assumption of all criteria to be of benefit-type is acceptable.

> ***Dominated Solution:*** *A feasible solution (or, alternative) X in a multiple criteria problem is a dominated solution compared to another feasible solution Y if Y is at least as good as (i.e., as preferred as) X for all criteria and is better than (i.e., preferred to) X for at least one criterion.*

Looking at Table 1.1 data, we find that alternative A_3 is a dominated solution compared to A_1 since all three criteria values of A_3 are less than the corresponding criteria values of A_1. Similarly, A_5 is also a dominated solution.

> ***Non-dominated Solution:*** *A non-dominated solution is a feasible solution, which is not dominated by any other feasible solution.*

That is, for a non-dominated solution an increase in the value of any one of its criteria is not possible without some decrease in the value of at least one other criterion. In Problem 1.1, to check if A_1 is non-dominated, it must be compared to both alternatives A_2 and A_4 (no need to compare with A_3 and A_5 as both are dominated). We find that A_1 is non-dominated because comparing A_1 and A_2, we see $50 < 75$, $-30 > -50$, and $25 > 7$; comparing A_1 and A_4, we see $50 > 20$, $-30 < -10$, and $25 > 18$. Similarly, we find that A_2 and A_4 are also non-dominated.

> ***Ideal Solution:*** *In a multiple-criteria problem, an Ideal Solution H is an artificial solution, each of whose criterion value is the maximum of that criterion's value for all potential feasible solutions.*

That is, the ideal solution consists of the upper bound (or the maximum) of the individual criterion values. The ideal solution is also known as the positive-ideal solution or the utopia solution. In Problem 1.1, the Ideal solution is $H = (75, -10, 25)$. This solution is imaginary and can be easily identified by selecting the maximum value under each C_j column. In all non-trivial multi-criteria problems, the ideal solution is infeasible.

> ***Anti-Ideal Solution:*** *The Anti-Ideal Solution L consists of the lower bound (or the minimum) of the individual criterion values.*

This solution is also known as the negative-ideal solution or the Nadir Solution. The pay-off table, such as Table 1.1, can be used to identify the anti-ideal solution. The minimum values in each column of the pay-off table constitute the anti-ideal solution. In Problem 1.1, the anti-ideal solution is $L = (15, -60, 5)$, which in this case is a real but dominated solution.

> ***Satisficing Solution:*** *A satisficing solution is a feasible solution that meets or exceeds the Decision-Maker's minimum expected level of achievement (or outcomes) for all criteria.*

In Problem 1.1, if the DM would be satisfied with any alternative with C_1 greater than or equal to ($\geq$) 35 and $C_2 \geq -40$, and $C_3 \geq 15$, then alternatives A_1 and A_3 are the satisficing solutions because only these two have satisfactory achievements for all three criteria. However, since A_3 is a dominated solution (compared to A_1), it would not be a candidate for a satisficing or preferred solution.

> **Best Compromise or Preferred Solution:** *The best compromise or preferred solution of a problem that has multiple, conflicting evaluation criteria is that feasible solution that has the criteria achievement values combination that is most preferred by the decision-maker.*

The preferred solution is identified by making trade-off decisions based on the DM's preferences. In Problem 1.1, the preferred solution will be one of the non-dominated solutions A_1, A_2, and A_4. As an example, using linear normalization described later in Chapter 4 and assuming that all three criteria are equally important to the DM, we can calculate the last column in Table 1.2 representing the weighted (weight of each criterion = 1/3) summation of criteria values for each feasible alternative and the preferred solution would be the one with maximum value. Based on Table 1.2 values, A_1 is the preferred solution.

TABLE 1.2
Normalized Achievement Values of Alternatives of Problem 1.1

Criterion →	C_1	C_2	C_3	
Weights →	$w_1 = 1/3$	$w_2 = 1/3$	$w_3 = 1/3$	**Weighted Row Total**
A_1	0.58	0.60	1.00	$(1/3*0.58) + (1/3*0.60) + (1/3*1.00) = 0.73$
A_2	1.00	0.20	0.10	0.43
A_3	Dominated			
A_4	0.08	1.00	0.65	0.58
A_5	Dominated			

1.6 CREATIVE PROBLEM-SOLVING

Many tools are available to help us solve problems easily and effectively. Problem-solving involves going through several steps, often iteratively. Understanding the underlying process or phenomenon that has given rise to the problem is often key to solving it. A decision aid, as we will use the term in this book, is a model, method, technique, or process designed to enhance the decision-making process.

Three common decision-making processes are:

> **Intuitive Decision-Making:** *This decision-making process has no detectable pattern, order, logic, or consistency.*

This approach is often associated with decision-making from the 'gut instinct' – the DM cannot clearly explain how they have arrived at the decision choice. Intuitive decision-making implies instinctive decisions without a conscious process for the development and evaluation of alternatives. Intuitive decisions usually have the following properties: (a) cannot be explained very well, (b) is a product of subconscious thought, and (c) private and internal to the DM.

One of the problems with intuitive decisions is that the DM cannot justify the choice made and, hence, it is often difficult to persuade others about its appropriateness. Note, however, that intuitive decision may not mean a complete absence of analysis. More likely, though, it may simply mean that the synthesis of analytical and other information is intuitive.

> **Programmed Decision-Making:** *This decision-making process is encoded in pre-specified decision steps – usually computerized.*

Programmed decisions are those that are made automatically, based on previous analysis or experience. For certain problem situations, it is possible to develop guidelines or instructions for decision-making based often on past analysis and experience. Usually, these problems are routine, and the DM has high confidence in the robustness of the solution approach prescribed for it.

> **Analytic Decision-Making:** *This decision-making process has the logic, pattern, or process that explains choices.*

Analytical approaches to decision-making involve structured analysis, often with the help of mathematical models, of the possible courses of action and their consequences. Analytic approaches provide structure and guidance for systematic thinking for solving difficult problems. There is considerable evidence to support the belief that systematic decision-making increases the probability of achieving a good outcome. There is also evidence that almost any systematic method of deciding, even though highly simplified, is better than none. Our objective in this book is to suggest decision aids that will systematize the decision-making. This is what we consider as creative problem-solving.

1.7   WHY ARE DECISIONS HARD TO MAKE?

In a typical decision situation, we find that several considerations make the decision-making process hard. The reasons could be any combination of,

- The complexity of the situation – that is, the DM faces many different issues that must be considered.

For example, in the new aircraft plant location example described earlier, the problem involves the consideration of the availability of sufficient land, the proximity of a general aviation airport, the transportation cost for parts and other subassemblies, the availability of trained or trainable production workers, taxes, etc. These must be considered while developing feasible alternative sites and their evaluation.

- The inherent uncertainty in the situation—estimates of each of the issues or factors delineated for evaluation purposes have associated uncertainties.

For the solid waste disposal problem, to what degree of certainty the future trash transportation cost can be estimated for decision-making?

- A DM may be interested in working toward optimizing multiple objectives, but progress in one direction may impede progress in others – that is, they are conflicting.

While comparing new aircraft plant location alternatives, the DM may find that the objective to minimize the total project cost is counter to the desirability of locations next to a general aviation airport (i.e., desirable locations have higher total cost) – the achievement of these two criteria are thus counter to each other.

- A problem may be difficult if different perspectives lead to different conclusions.

For the new aircraft plant location example, a desire to locate the assembly plant in the state where the parts factory is located might lead to one selection while a desire to locate in a 'right to work' state may divert the analysis to another direction.

1.8   THE DECISION-MAKING PROCESS

Decision-making usually is a process consisting of the following steps:

Step 1. *Identify and define the decision situation or the decision problem.*

This is often the key step in problem-solving. Consider a very simple situation: you are feeling feverish and go to your primary care physician for a checkout. Your doctor can just look at your body temperature and tell you to take an aspirin tablet and rest. But the doctor knows that the raised body temperature may just be a symptom that is caused by something else such as an infection. The doctor will therefore check your ear, nose, tonsils, and do blood and other tests to determine what is causing the fever. After these tests, if they conclude you have a sinus infection, then the doctor probably will prescribe an antibiotic as well as other medication. In other words, the doctor had to study and analyze before they identified the actual problem to be solved.

Step 2. *Identify an appropriate problem model, the judgment criteria to be used, and the data needs.*

Once we have identified and defined the problem, we then must decide on the model or approach to be used to identify alternatives, decide on the judgment or evaluation criteria, and identify the data needs. Note, however, that a model only approximates the reality, requiring us to make some assumptions to simplify the problem to make it solvable. We must balance between solvability and retention of reality. In Chapter 2 some tools for systematic problem analysis have been described.

Step 3. *Identify the feasible alternatives.*

In Chapter 3, some methods for developing alternatives and criteria are described.

Step 4. *Evaluate the alternatives and choose the best alternative.*

In Chapters 5 and 6, methods for evaluating and selecting the preferred alternative are described.

Step 5. *Perform sensitivity or 'what if' analysis, if necessary.*

This step allows us to assess the robustness of the preferred alternative by performing a 'what if' analysis by varying some data and/or model assumptions.

1.9 EXPLANATION OF THE MATHEMATICAL NOTATIONS USED

We review here the mathematical notations used in the book. This will make it easier to understand the tools and techniques described in the later chapters.

a. Symbol '*' has been used in equations to indicate multiplication

b. X_i with $i = 1, 2, 3$ means X_1, X_2, X_3

c. $\begin{bmatrix} X_{11} & X_{12} & X_{13} \\ X_{21} & X_{22} & X_{23} \end{bmatrix}$ means a matrix with 2 rows ($i = 1,2$) and

   3 columns ($j = 1,2,3$)

d. X_{ij} means X value in row i and column j of a matrix

e. $\sum_{i=1}^{i=4} X_i$ means $X_1 + X_2 + X_3 + X_4$

f. X^3 means X raised to the power of 3 or $X*X*X$

g. $|X|$ means the absolute value of X, that is $|-X| = X$ and $|X| = X$

h. $\prod_{i=1}^{i=3} X_i$ means $X_1*X_2*X_3$

i. $\sqrt[3]{X_1*X_2*X_3*X_4}$ means $(X_1*X_2*X_3*X_4)^{1/3}$

1.10 CLASSIFICATION OF DECISION PROBLEMS

Table 1.3 provides a general classification of decision problems and the solution methods that can be used for solving problems with a single evaluation criterion. Table 1.4 shows the methods that can be used for solving problems with multiple evaluation criteria. These tables can be used as a guide for selecting a solution approach for a specific problem. Note that often multiple solution methods can be applied, and the selection of the method to be applied may be based on the DM's comfort level in applying a method and the amount or type of data available.

TABLE 1.3
Classification of Decision Methods with Single Criterion

Without Uncertainty
1. Single Criterion Optimization Method

With Uncertainty
1. Laplace Method
2. Max-Min Method
3. Min-Max Regret Method
4. Expected Value Method
5. Decision Tree Method

TABLE 1.4
Classification of Decision Methods with Multiple Criteria

Methods Requiring No Decision-Maker Preference Information
1. Min-Max Regret Method
2. Compromise Programming Method

Methods Requiring Decision-Maker Preference Information
1. Effective Value Method
2. TOPSIS Method
3. AHP Method

2 Tools for Problem Analysis

2.1 INTRODUCTION

An important step in any problem solving is analyzing the perceived problem for a better and clearer understanding of it. In this chapter, we will explain some commonly used tools for problem analysis. Not all these tools will be applicable nor needed to be used in every problem the Decision-Maker (DM) would encounter. Ishikawa (1982) has more detailed discussions about some of the analysis tools described later.

2.2 FLOWCHART

A process is defined as a series of actions taken to produce a specific output (e.g., an ordering process) and a process flowchart is a graphical illustration of the process that shows the interrelationships among the tasks as well as the tasks themselves. A process flowchart is a visual communication tool used for process definition and understanding. It can be used to identify critical points in the process where problems occur and where the DM may need more information. In a specific problem solving situation, multiple flowcharts may be created: one for the initial problem and others for the proposed solution alternatives. Creating a process flowchart consists of the following steps.

Step 1. *Process Identification and Clarification.*

In identifying the process to be studied, DM needs to be as specific as possible to the area to be improved or corrected. Usually, it is better to start with a broad area of the problem and then focus on specific steps for more details as needed. The identification process clarifies what the process is supposed to accomplish, and its start and endpoints. The questions to be asked in this step are:

a. Who is the process owner? The process owner is that individual who has the authority or ability to make changes.
b. Who are the customers? Customers are the recipient of the outputs of the process. They can be internal and/or external to the organization.
c. Who are the suppliers? Suppliers are the source of information, material, and people entering the process. A supplier can be internal and/or external.
d. What are the process inputs, i.e., what information or material flows into the process at the start point or later?

DOI: 10.1201/9781003172291-2

e. What are the process outputs, i.e., what information or material leaves the process endpoint?

Step 2. *Determine the Level of Detail.*

The level of detail may depend on where in the organization the flowchart is being created. For example, the level of detail of the problem for an organizational marketing process would be different from the level of detail when a regional manager is looking at the marketing of a specific product portfolio. At each level of analysis, the DM should assure that the sufficiency of details matches the reason for the analysis. Too little detail will miss in identifying the problem points, and too much detail will create a haystack where the problem points will be difficult to identify. If the flowchart does not help the DM to understand the process well enough to improve it or correct the problem, then more details need to be added. More details may be needed in only a few process steps, while other steps may be looked at more broadly.

Step 3. *List the Process Steps.*

List all steps in chronological order. The first process flowchart should reflect the process as it currently is. During the interpretation phase (alternative development phase) you may redesign the process as it should be. Ask what is moving through the process. Some examples are material (parts, equipment), information (ideas, communication), and people.

Step 4. *Create the Flowchart.*

Flowcharts are created by using symbols. Symbols are helpful; they help people see the type of activity occurring. Common symbols are:

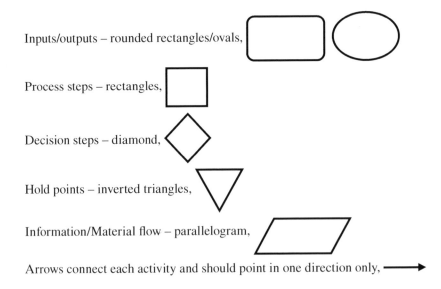

Inputs/outputs – rounded rectangles/ovals,

Process steps – rectangles,

Decision steps – diamond,

Hold points – inverted triangles,

Information/Material flow – parallelogram,

Arrows connect each activity and should point in one direction only, ⟶

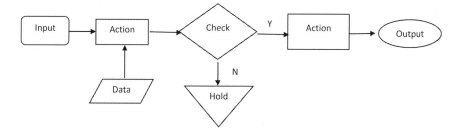

FIGURE 2.1 Basic structure of a flowchart.

Multiple flows occur at decision points. If a condition is met, then take the next step. If it is not met, then take an alternate step. A basic structure of a process flowchart is shown in Figure 2.1.

2.2.1 An Example of a Process and Its Flowchart

As the Sales Manager, you would like to analyze how online sales orders are handled to see if the process can be improved. Develop an as-is process flowchart.

Step 1.
Process Owner – You (Sales Manager).
Customers – Customers submitting the order online.
Suppliers – Shipping department.
Inputs – Individual purchase orders, warehouse data.
Outputs – Shipment date confirmation to customer or out-of-stock note to customer if the item would not be reordered by the warehouse.
Step 2. You have decided to view the process at a macro level, considering only broad process elements. A more detailed analysis will be done at the alternative development phase.
Step 3. The initial process steps are incoming customer orders, order completeness check, availability of the ordered items, shipment date confirmation, and out-of-stock notification.
Step 4. Develop the process flowchart of how sales orders are handled now. Figure 2.2 shows the developed chart.

2.3 CHECK SHEET

Check sheets are tools for collecting and organizing facts and data, usually at the point where the data is generated. The collected data can be quantitative or qualitative, historical or observations as they happen. There are many different types of check sheets depending on the use. Three common types of check sheets are recording sheets, tally sheets, and measles charts. A recording sheet is used to collect measured data. Counted data is collected by making tick marks in a tally sheet. A measles chart is used to collect locational data. Generally, measles charts are pictures, illustrations, or maps onto which data is collected.

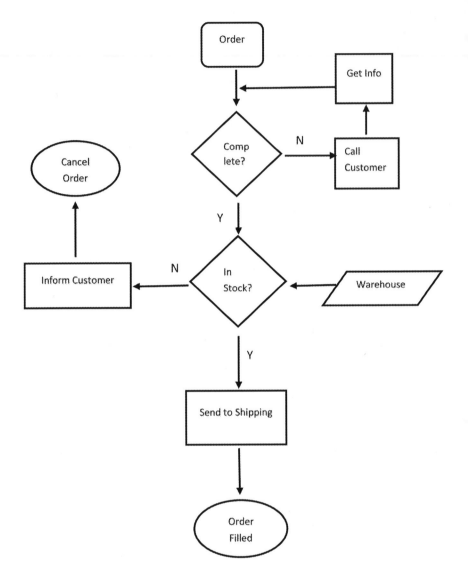

FIGURE 2.2 Process flowchart of sales order handling.

2.3.1 Steps in Creating Check Sheets

Step 1. *Define events or conditions to be observed.*

Clearly define and have a common understanding of what is to be observed. Examples of events or conditions are – what is considered as a defective product, what does slow speed mean, what is incomplete, etc.

Step 2. *Plan data collection.*

How, where, and the period over which data is to be collected are decided at this step. Questions that need to be answered are – is the data to be from samples or the entire population; at which process step will the data be collected (at shipping, at machine #1, sales orders received online, etc.); how will the data be collected (from company's existing records or to be observed, automated scanners at shipping, temperature checkpoint at building entrance, etc.); and over what period will the data be collected (for the next 3 months, 12 weeks, from 8:00 am to 5:00 pm, etc.).

Step 3. *Create the check sheet.*

Design a check sheet template that is clear, complete, and easy to use. Ensure that the template includes the data collection location, data to be collected, any necessary data identifier, name of the event for which data is to be collected, times or dates when data will be collected, etc.

Step 4. *Collect data.*

Collect and record the data consistently and accurately. The data collection and recording can be completely or partially computer-based.

2.3.2   AN EXAMPLE OF A CHECK SHEET

We want to investigate the types of defects found at the final inspection station before the product is sent to the warehouse for shipping.

Step 1. Defects to be observed – motor failure (of any kind), painting defect (of any type), electrical circuit defect/failure, and hose leak.
Step 2. Data will be collected as follows:
   • Test all finished products
   • Inspect at final inspection station before warehousing
   • Use test results from an automated inspection machine
   • Collect daily counts for each of the two daily shifts over 1 week (5 days)
Steps 3 and 4. An example of collected data and entered in the check sheet template is shown in Table 2.1. Note: a defective unit may have multiple failures.

An example of a filled tally sheet is shown in Table 2.2.

A measles chart of painting blemishes on a part may look like in Figure 2.3 (after 1 week of data collection). The numbers are counts of observed blemishes in a specific area of the part.

2.4   HISTOGRAM

A histogram is a bar chart for variable data that are grouped into classes. Histograms can be used to show a picture of data distribution, including the center, spread, and shape, or to compare data distributions to specification requirements, or to compare

TABLE 2.1

Recording Sheet of Defects Observed at Final Inspection Station

Day	Mon, 10/05		Tue, 10/06		Wed, 10/07		Thu, 10/08		Fri, 10/09		Row Total
Shift #	1	2	1	2	1	2	1	2	1	2	Total
Total Checked, #/shift	950	500	955	505	960	498	965	510	966	509	7,318
Total Defective, #/shift	7	2	0	1	4	0	1	1	0	0	16
Motor Failure, #	2				1						3
Painting Defect, #	5	2			1		1				9
Electrical Circuit Failure, #	1			1	2						4
Hose Leak, #								1			1
Column Total, # of failures	8	2	0	1	4	0	1	1	0	0	17

TABLE 2.2

Tally Sheet of Defects Observed during Shift 1 at Final Inspection Station

Day	Mon, 10/05	Tue, 10/06	Wed, 10/07	Thu, 10/08	Fri, 10/09	Row Total
Shift #	1	1	1	1	1	Total
Total Checked, #/shift	950	955	960	965	966	7,318
Total Defective, #/shift	┼┼┼┼ | |		| | | |	|		12
Motor Failure, #	| |		|	|		4
Painting Defect, #	┼┼┼┼		|	|		7
Elect. Circuit Failure, #	|		| |			3
Hose Leak, #						0
Column Total, # of failures	8	0	4	2	0	14

data distributions for different processes, times, and equipment. Steps in creating a histogram are:

Step 1. *Decide on the number of observations. Collect data.*

Count the number of observations, determine the data range, and the two endpoints (smallest, largest).

Step 2. *Decide on the number of classes or bars for the histogram.*

The number of classes or bars to be used depend on the number of data points and the level of details wanted. An often-suggested rule of thumb for the number of classes or bars is:

# of observations 20–50, # of classes 5–7
# of observations 50–100, # of classes 6–10

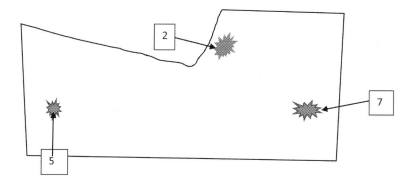

FIGURE 2.3 A measles chart for painting blemishes in a part.

# of observations 100–250, # of classes 7–12
# of observations 250+, # of classes 10–20

Step 3. *Determine class/bar boundaries.*

Select the class/bar boundaries so that they are non-overlapping, convenient, and appropriate. Count the number of observations in each class.

Step 4. *Draw the histogram.*

Draw horizontal/vertical axes; scale and label each axis. The vertical axis should be able to accommodate the highest frequency class. Draw the vertical bars for each class. The height of the bars equals the frequency for the class. The horizontal axis represents the data range and shows all bars.

2.4.1 An Example of a Histogram

The sales department wants to know how long it takes to ship an order after the order has been received. The sales order data for the most recent one complete month is to be used.

Step 1. Shipment can take 1–15 days after an order is received and the total number of shipments in any 1 day maybe, on average, up to 30. Collected data for the number of days to ship after an order is received and the number of shipments on a day after the order is received is shown in Table 2.3.

Step 2. It may take up to 15 days to ship, so we decide to use five classes.

TABLE 2.3

Shipment Data for the Histogram Example Problem

# of Days to Ship	1	2	3	4	5	6	7	8	9	10	11	12	13	14	15
# of Shipments	11	30	11	2	9	16	0	12	10	13	12	19	18	4	7

Step 3. Bar boundaries to be used are 1–3, 4–6, 7–9, 10–12, and 13–15 days.

Step 4. Figure 2.4 shows the histogram representation of the same shipment data.

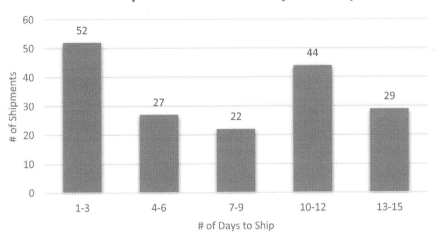

FIGURE 2.4   Histogram of # of shipments vs # of days to ship.

2.5   PARETO CHART

A Pareto chart is a graphical tool with both bars and a line graph, where descending ordered bars represent categorized values, and the cumulative total is represented by the line. The order of the bars reflects the importance or costs of the causes or categories associated with the bars. It is used to identify the most frequent defects, complaints, or any other factor that is countable and categorizable starting from the most frequent to the least frequent to focus efforts on the factors that produce the greatest impact. The Pareto principle states that a few of the causes result in the major number of defects, complaints, costs, etc. This principle is also referred to as the 80–20 principle, that is about 80% of the factors result from about 20% of the causes. The Pareto chart can be used to identify the few causes where corrective action focus should be on. Steps in creating Pareto charts are as follows.

Step 1. *Define the problem.*

Before creating the charts, decide on the problems, causes, and defect categories. Clearly define and list what is to be studied; that is, which problem types, causes, items, etc. are to be analyzed. Decide how these items are to be measured (frequency of occurrence, cost, time taken, number, etc.) – data must be countable and be categorizable. Set the time over which data would be collected (observed or from company records).

Step 2. *Gather data.*

Collect (new observations) or compile (from company records) data over the period already decided on using a well-structured data collection system.

Step 3. *Create Pareto chart.*

Decide first on the number of data categories (like that used in bar charts). Then,

- Organize data into the selected categories.
- List the categories in descending order of the measure of comparison (number, frequencies, cost, profit, etc.).
- Compute the cumulative totals by adding the categories in the order listed.
- Draw, label, and scale the axes. Divide the horizontal axis into equally spaced intervals – write the category labels starting with highest on the left. The left vertical axis is to show the category data/numbers (frequency, time, or cost). The right vertical axis is to show a percentage scale of the cumulative total percentages (the cumulative total percentage should add to 100 percent).
- Draw the bars for each category. The bars should be in descending order of most to least.
- Draw a line graph of the cumulative totals. The first point on the line graph should line up with the first bar.

2.5.1 AN EXAMPLE OF A PARETO CHART

A company has ten factories. For 2 weeks, through a customer survey, it has compiled the number of customers who were unsatisfied with the quality of the most popular product. It wishes to know which factories produce the most defective items so that corrective actions can be taken to reduce defectives.

Step 1. We want to determine which factory's products result in the most unsatisfaction so that corrective actions could be taken in those factories.
Step 2. The collected data is shown in Table 2.4.
Step 3. Categories – We will use factories as categories.

Pareto chart is shown in Figure 2.5 – the gray bars are factory data, and the black line is the cumulative frequency percentages. It appears that Factories 2, 6, and 10 account for about 80% of the unsatisfied customers and should be the focus of further

---

TABLE 2.4

Number of Unsatisfied Customers in Different Factories

Factory #	1	2	3	4	5	6	7	8	9	10
# of Unsatisfied Customers	5	45	5	2	3	35	2	10	3	20

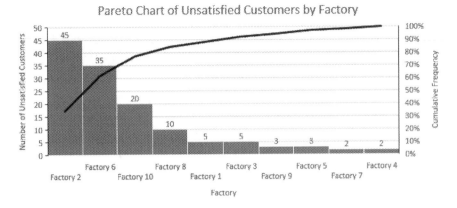

FIGURE 2.5   Pareto chart of unsatisfied customers by factory.

analysis (perhaps using C–E diagram analysis) to identify what corrective actions need to be taken.

2.6   CAUSE-AND-EFFECT (C–E) DIAGRAM

A C–E diagram (also called a fishbone diagram or Ishikawa diagram) is a graphical technique that lists and organizes possible problem causes or contributors to a specific problem or effect. It illustrates the relationship between a given problem or effect and all the factors that influence or cause the effect; it is also helpful in assessing additional causes. We can then isolate the most likely causes or contributors and focus on corrective actions for those causes. The '5 whys' technique is often used to explore the C–E relationships. The goal of using '5 whys' is to determine the root cause of a defect or effect by repeating the question 'why'. Each answer forms the basis for the next 'why'. The '5' in the name derives from the fact that often about 5 iterations are needed to get to the root cause. However, five iterations may not be needed in all problems. Steps in performing a C–E diagram analysis are:

Step 1. *Clearly define the problem to be solved.*

Agree on the problem statement (also referred to as the effect). This is something we want to improve and control. Do not define the problem as a solution statement such as Need More Auditors.

Step 2. *Decide on the major categories.*

Agree on the major categories of causes of the problem (written as branches from the main arrow) – usually 4–6. The causes will depend on the problem being solved. Commonly used first-level causes are machine (equipment), method/policy, manpower/people, systems, skill, and cost.

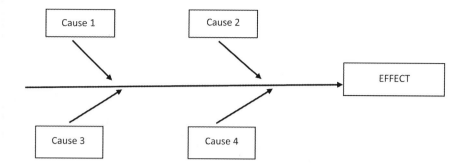

FIGURE 2.6   A Simple cause and effect diagram.

Step 3. *Ask 'why' iteratively.*

Brainstorm all the possible causes of the problem. Ask 'Why does this happen?' Each major cause will form a branch. For each cause, ask again 'why' to branch off. Repeat this a few times until the root cause(s) is(are) identified. A root cause is the one that can be addressed to prevent future problems. A simple C–E diagram may look like Figure 2.6.

2.6.1   A C–E DIAGRAM ANALYSIS EXAMPLE

It is a concern in an accounting firm that supervisors are finding mistakes being made in preparing the IRS 1040 form. We need to find out how that can be fixed.

Step 1. The problem (effect) – The number of mistakes in completed Form 1040 is high.
Step 2. After brainstorming by the supervisors, the identified major causes are people, process, equipment, and environment.
Step 3. After repeating 'why' one more time, the C–E diagram is created as shown in Figure 2.7. We can continue to iterate one or two more times until we can identify a few root causes for which corrective actions can be taken. For example, we can ask 'why' workstation causes errors. The responses (e.g., table height, chair, etc.) can then be branches off the 'workstation' cause.

2.7   AFFINITY DIAGRAM

An affinity diagram is a tool used to help identify creative or nontraditional solutions to a problem by organizing large amounts of information by grouping ideas based on key relationships between each item. It is usually a team activity that is preceded by idea generation during brainstorms. An affinity diagram helps in understanding a complex problem or issue, as well as identifying potential causes of a problem, and groups the generated ideas according to their affinity or similarity. It may be used in new product/service development or in situations that seem confusing/disorganized or when the team has incomplete knowledge of the area being investigated. The ideas are usually written on sticky notes as they are generated so that they can be moved

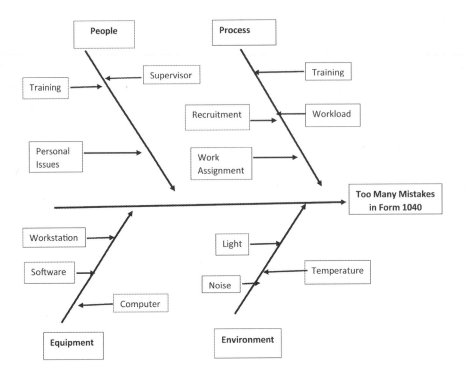

FIGURE 2.7   C–E diagram of IRS form 1040 mistakes.

around later to find clusters of similar ideas. Once clusters have been identified, the clusters may be used to create a C–E diagram. The common steps in developing an affinity diagram are as follows.

   Step 1. *Define the problem. Generate and record each idea on a separate sticky note or computer (connected to a projector).*

At this stage, after defining the problem clearly, each member starts contributing ideas on sticky notes. Spread notes of ideas randomly on a large work surface so that all notes are visible to everyone. The entire team is involved in this and all steps.

   Step 2. *Find common themes for clustering.*

Look for ideas that seem to be related in some way and place them side by side. Look for relationships between individual ideas and have members simultaneously sort the ideas in silence to encourage unconventional thinking. As members sort the ideas, they should ask themselves if the idea in the note is like any other note and belongs to the same group or cluster. If not, make it the first sticky in the second group. This process of sorting is repeated until all notes have been grouped and all members are agreeable to the groupings. The common themes should not be predefined, rather they should emerge organically as part of the process. Eventually, 3–10 related groups will emerge. Name each one of them based on the common theme.

Step 3. *Create 'supergroups' if appropriate.*

Sometimes it may be possible to group some clusters into a supergroup. Once the notes have been sorted into groups or supergroups, the team may sort supergroups into subgroups for easier management and analysis.

Step 4. *Create Headers.*

Members decide on phrases that best describe the grouping. The header may even be the note capturing the main idea in that group or cluster. These phrases are the headers for that group. Each cluster represents a major concern or component of the problem.

2.7.1 AN AFFINITY DIAGRAM EXAMPLE

Consider the IRS 1040 Form mistakes problem described in the C–E example. We want to determine the causes of this problem.

Steps 1 and 2. The ideas generated by the investigative team are noted on a
   computer (sticky notes here) and then clustered around common themes.
   Group 1: Training of form preparers, appropriateness of supervision, personal issues (such as childcare, commute, etc.) of preparers.
   Group 2: Workstation (is it adjustable, enough desk space, etc.), software (is
   it up to date, easy to use, backs up automatically, etc.), computer (is it
   fast enough, large enough screen size, etc.).
   Group 3: Temperature, light, background noise.
   Group 4: Is work assignment appropriate to preparer's ability, is the work-
   load appropriate, is the recruitment process for preparers appropriate,
   is the training appropriate.
Step 3. No supergroups.
Step 4. We will use People, Equipment, Environment, and Process as headers
   for the four groups and draw the affinity diagram as shown in Figure 2.8.

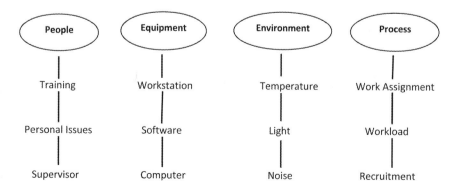

FIGURE 2.8 Affinity diagram for the example problem.

2.8   TREE DIAGRAM

A tree diagram (or systematic diagram) is a flowchart-like diagram that can be used to map the various outcomes from a series of decisions. It is used to break down broad categories into finer and finer levels of detail. It allows the DM to weigh possible decisions against one another based on their costs, probabilities, and benefits. They are used to visualize all possible outcomes and decision points. When working on a broad goal, a tree diagram helps in focusing on specific tasks. It can be used to plan and implement a detailed solution in an orderly manner. The tree diagram also serves to visually break down a complex issue or goal.

A decision tree starts with a root node representing the problem to be solved. It is then followed by two different types of nodes: decision nodes and chance nodes. Finally, it terminates in end nodes which represent the outcomes. A decision node represents a decision point, and a chance node represents the probabilities (or chances) of a specific outcome. The nodes are connected through arrows or lines or branches representing flows from one node to another node. Developing a tree diagram involves only a few steps.

Step 1. *Define the problem and collect data (information).*

This step may follow a previous analysis of the problem using any of the techniques described earlier. We start with the root node that is a goal statement, which is the broad goal or problem that can be systematically broken down into actions or means. It may be a key issue or most critical header from an affinity diagram, or a root cause/effect from a C–E diagram, or a team-identified key issue (after brainstorming all possible issues).

Step 2. *Add branches.*

Add branches for all possible alternatives or outcomes from the root node. A line or arrow represents each branch (alternative or outcome) and connects to a decision node or a chance node. To identify the branches, we ask what do we need to address to achieve this goal or solve this problem?

Step 3. *Continue adding nodes and branches.*

For each decision node or a chance node, add branches for each possible outcome. Continue adding nodes/branches till every question has been resolved and an outcome has been reached. The diagram is complete when we cannot create any more levels.

Step 4. *Add end nodes and review the completed tree.*

The node where a final solution or outcome has been reached becomes an endnote. Finally, a review should involve to check for sufficiency and completeness.

2.8.1 A TREE DIAGRAM EXAMPLE

Consider a situation where the management is trying to decide if technology A or technology B should be selected for a new factory to be established. A team of experts in various areas has been selected to work on this.

> Step 1. The root node will be 'Find the technology to be selected'.
> Steps 2 and 3. The tree diagram in Figure 2.9 shows nodes and branches as developed.

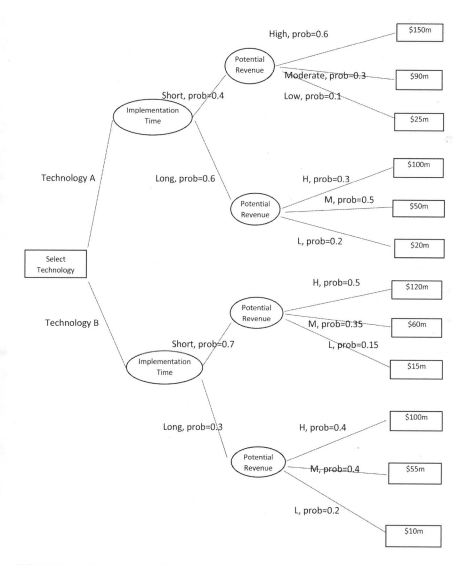

FIGURE 2.9 Tree diagram of technology selection problem.

Step 4. The end nodes show the potential revenue.

Usually, a collapsing of the tree is done starting from endnotes to the root node by following each branch. In the example problem, we compute the expected revenue of each technology. We compute expected revenue because the tree has chance nodes with associated probabilities involved. Expected revenue for each technology is calculated as follows:

- Expected revenue of Technology A = 0.4*(0.6*$150 + 0.3*$90 + 0.1*$25) + 0.6*(0.3*$100 + 0.5*$50 + 0.2*$20) = 0.4*$118.5 + 0.6*$59 = $82.8 m
- Expected revenue of Technology B = 0.7*(0.5*$120 + 0.35*$60 + 0.15*$15) + 0.3*(0.4*$100 + 0.4*$55 + 0.2*$10) = 0.7*$83.25 + 0.3*$64 = $77.48

So, we will select Technology A for implementation or additional analysis.

2.9   PROBLEM ANALYSIS METHOD SELECTION ROADMAP

Table 2.5 can be used as a rough guide to decide which tool is to be used for analysis in a particular problem. The choice of a specific tool will depend on the type of data available and the analysis purpose.

TABLE 2.5
Problem Analysis Tool Selection Decision Table

	Problem Analysis Tool						
Purpose	Flowchart	Check Sheet	Histogram	Pareto Chart	Cause and Effect Diagram	Affinity Diagram	Tree Diagram
Describe Problem	√	√	√	√	√	√	√
Identify Root Cause	√	√	√	√	√	√	√
Plan Action	√			√	√		√

3 Developing Alternatives and Criteria

3.1 OVERVIEW

Decision-making consists of three actions: identifying the problem (which sometimes is also referred to as finding the 'pain point' to be addressed), finding possible courses of action or alternatives, and choosing among the courses of action. In Chapter 2 we have discussed some tools for identifying and defining the problem to be solved. What we must be careful about in the problem identification step is not to be led astray by focusing on the symptoms of a problem rather than the causes of the symptoms. In this chapter, we discuss some approaches to the development of alternatives and evaluation criteria for a decision problem. The process for alternative development and criteria development is somewhat similar. So, while we will discuss much on alternative development, most of the approaches can also be used for evaluation criteria development.

3.2 CREATIVITY BLOCKS TO ALTERNATIVES AND EVALUATION CRITERIA DEVELOPMENT

As indicated previously, a decision problem implies the existence of more than one alternative. Depending on the problem setting, decision-making environment, and the complexity of the relationships of factors to be considered, the process of developing decision alternatives may be explicit or implicit. In the case of explicit alternative development, all potential alternatives should be identified. A failure to consider a large enough set of alternatives may result in the exclusion of a better potential 'best' alternative that was not considered. In the case of implicit alternative development, the choice of alternatives is limited to those implied through the mathematical relationships used. In this book, we will focus on problems where explicit development of alternatives is possible.

Developing alternatives and evaluation criteria is comparable to a creative process. Thus, it helps to have an idea of what blocks or fosters creativity. In the literature on the theory and practice of creativity, the following reasons for creativity blocks are often highlighted.

- *Perceptual blocks* that stop us from observing and/or understanding the problem. Examples of this type of creativity block are the inability to see the problem from various points of view, limiting the problem area too narrowly, considering too many concerns, and the sheer volume of information (that may keep us from seeing new and promising ideas).

DOI: 10.1201/9781003172291-3

- *Emotional or mental blocks* have to do with personality traits that affect how we behave. Examples of this type of block are anxiety with taking a risk that reduces the ability to ponder fresh ideas, inability to give enough time to contemplate, and judging ideas too soon that may block new vision and new ideas.
- *Cultural and Organizational blocks* have to do with societal and organizational norms and practices. Examples of these types of blocks are inertia for the status quo, societal and cultural notions of acceptable behavior, work environment, tolerance or (even) encouragement for out-of-the-box thinking, and organizational control structure.

3.3   TECHNIQUES FOR ENHANCING CREATIVITY

The question now is how to minimize the impacts of these blocks and encourage creativity? Practitioners of creativity boosting suggest the following techniques for enhancing creativity.

3.3.1   Idea Checklist

Various suggestions for developing a list of ideas through the systematic use of action or descriptive verbs are summarized here. These approaches can be used (often by brainstorming) to develop solution alternatives and evaluation criteria.

- Adapt what we already have for a different problem.
  - What else is like this?
  - Is there something like it, but in a different context?
  - What other ideas does it suggest?
- Modify, lessen, amplify, streamline existing
  - Purpose
  - Input
  - Process
  - Value
  - Design
  - Use
- Substitute with something else. Ask: what other ... can be used?
  - Component
  - Process
  - Input
  - Approach
- Rearrange
  - Components
  - Pattern
  - Layout
  - Process steps
  - Sequence

- Put to other use. How and where else can it be used?
  - Use as is in another context (product, process, organization, market, etc.)
  - Use with modification in another context
- Combine or eliminate existing
  - Components
  - Processes
  - Steps
  - Ideas
  - Purposes

3.3.2 Brainstorming

Brainstorming is a group activity. In this approach, two or more people go around each other and contribute ideas until enough new ideas have been generated or no new idea is forthcoming. Participants may build on or modify ideas already contributed. The focus of brainstorming is to develop as many ideas as possible around a central theme without any critical assessment of those ideas at the generating stage. To be effective, a brainstorming session must ensure that:

- it is a limited size group activity with members from diverse backgrounds.
- no criticism, discussion, or evaluation takes place while ideas are being generated (i.e., these sessions should be noncritical at the idea generation stage).
- all participants are encouraged to think broadly, even come up with wild ideas. Thus, sometimes the supervisor is not allowed to be part of the group generating ideas.
- no participant or idea is cut off while ideas are still being contributed by participants.
- silence is maintained so as not to interfere with the spontaneity of the process and limit the flow of ideas.
- all ideas are recorded and kept visible.
- combining, modifying, expanding other ideas are acceptable.
- building-off of someone else's idea is encouraged.

In generating ideas, participants can use some of the approaches discussed under the problem-solving checklist. Assuming that the problem to be investigated has been defined clearly, the steps followed in brainstorming are:

Step 1. *Form a team and identify a facilitator.*

A limited size team with members from diverse backgrounds and expertise should be formed and a facilitator to lead the team in idea generation should be identified. All the team members should be in a meeting room (or be present virtually). The facilitator will need to establish definitions of key terms so that everyone has the same understanding, and compile any additional information participants will need, and send it to them or present it before and during the session. Next, the context of the problem question should be made clear. Giving some thought to the problem to be solved is very important. The problem question should be clear and prompt participants to think of solutions. Often the question can be formed as 'why' or 'how'.

Step 2. *Generate ideas.*

To assure that no participant is sidelined, the facilitator may use round-robin brainstorming, where each participant is asked, in turn, to contribute an idea. The idea generation will continue with each contributing one idea at a time until their turn comes next. A member's idea may be a new one or a modification of a previously identified one. No criticism is allowed during idea generation. Members are to be encouraged to think 'out-of-the-box'. After going around several times, if no new idea is forthcoming, the process is stopped.

Step 3. *Sort ideas.*

At this step of the brainstorming session, sort through the ideas and remove any obvious duplicates. While various steps for successful brainstorming have been suggested, keeping things simple appears to work well.

3.3.3   SWOT Analysis

SWOT analysis is usually a group activity but can also be performed individually. SWOT stands for Strengths, Weaknesses, Opportunities, and Threats. Thus, a SWOT analysis is used to assess these four aspects of the organization, product, or process to create new alternatives. In SWOT analysis the participants should preferably represent various parts of the organization and the expertise available in the organization. Processes such as brainstorming can be used for idea generation.

Strengths and weaknesses are internal or are the factors over which the organization has some control and can make changes to. Strengths are the things that the organization does well. These may include:

- Human, physical, and financial resources of the organization
- Activities, processes, systems in use in the organization
- Things that the organization does well and separates it from the competition
- Copyrights, patents, trademarks, proprietary technology/intellectual assets that the organization owns (or has a license to)
- Reputation in the community or the market

Weaknesses are the things the organization lacks in or does not do well. These may include:

- Things that the organization does poorly compared to its competitors
- Human, physical, financial resources the organization lacks in or has less of
- Lack of intellectual assets
- Systemic deficiencies in physical structures, processes/technologies used, and organizational checks and balances

Opportunities and threats are external to the organization and it usually has no control over these. Opportunities may include:

- New markets for current products and services or new products and services
- Little or modest current competition, the potential for product/service portfolio expansion
- Enhancing market reputation

Threats may include:

- Local, state, national, global, or sector economy trends
- Consumer, societal, cultural trends
- Weather events, population demography changes
- Technology changes, climate change, regulatory changes

Table 3.1 shows one template for grouping ideas from a SWOT analysis. If the analysis is performed as a group activity, approaches from the brainstorming process can be used during idea generation. Once the analysis is completed, the organization then can select which opportunities to take advantage of, which strength is to be further strengthened, which weakness can be addressed to minimize its impact, and which threat is to be further protected against.

TABLE 3.1
SWOT Analysis Template

Strength	Opportunity
• Modern factories	• Add a new product line
•	•
•	•
Weakness	**Threat**
• Limited product range	• Lack of available skilled labor pool
•	•
•	•

3.3.4 Nominal Group Technique (NGT)

NGT is like brainstorming and its process steps are similar. It works in a more structured way, with one member acting as the facilitator. It works best with small groups (preferably less than 8–10 participants). The NGT steps are:

Step 1. *The facilitator, after describing the problem to be solved, asks all participants to brainstorm (silently) and have them write down their ideas privately (i.e., not shared or discussed with other team members).*

Step 2. *The facilitator then asks each member, by going around the members in a round-robin fashion, to share one idea at a time (no criticism is allowed at this step).*

Step 3. *After all the ideas have been shared (and recorded on a chart or board), they are discussed by the group. The purpose of this discussion is to clarify the ideas presented and reducing overlap among them by combining where feasible.*

Step 4. *Each participant independently (privately) develops a ranking of all the ideas.*

Step 5. *Finally, the facilitator combines the individual rankings mathematically (usually by calculating a simple average) to generate a group ranking.*

3.3.5 DELPHI TECHNIQUE

Delphi technique is also a group process but is more suitable for a problem where there is a lack of hard data or is costly to obtain, and for situations needing predictions of future status. It involves a group of experts who participate anonymously. The participants should not be in one location and are not told who the other participants are. The end outcome is the most reliable consensus opinion of the group of experts. The experts are those who have relevant knowledge and experience about various aspects of the problem being studied. They can be from within or outside of the organization. The expert opinions are collected through a series of questionnaires (or, surveys) with controlled feedback from the facilitator (usually in the form of statistical metrics such as mean, mode, standard deviation, etc.) indicating 'group response'. The expert responses are always anonymous, that is the experts never know who made what response. After a relatively few rounds of the opinion-feedback loop, the consensus opinion emerges. The common steps of the Delphi technique are:

Step 1. *Identify a facilitator, define the problem, and identify the expert panel.*

The process starts with identifying a facilitator who identifies the experts to be used. The experts remain, throughout the process and at the end of the process, anonymous to each other. The facilitator then prepares a clear and complete definition of the problem to be addressed by the experts.

Step 2. *The facilitator starts the gathering of expert responses.*

The facilitator sends a survey or a poll to the experts inquiring about some general questions for identifying the experts' understanding and their perspectives about future events, markets, competitions, policies, etc. surrounding the issue to be forecasted. These questions are designed to pick up the comprehension of the specialists and their perspectives about the issue under consideration. The nonrelevant and overlapping responses are deleted to identify the basic perspectives.

Step 3. *Continue with the next rounds of questions.*

In subsequent rounds of questions (through new surveys or polls), the facilitator digs down on the common prior responses (where most experts agree) to further explain these issues. After each round of receiving expert responses, the facilitator deletes nonrelevant and overlapping responses. Feedbacks are sent to the experts that provide statistical summaries of responses. These summary results are used by the experts in responding to the new round of questions. The process of expert responses, summary feedback, and new responses from experts is repeated as many times as necessary (usually, 3–5 rounds).

Step 4. *Identify the future action item.*

It is expected that the experts will have arrived at a consensus and the facilitator will have a clearer perspective for the future event. From the final round responses, equally-weighted averages of the identified scenarios, forecasts, are used to group rank order the potential outcomes.

4 Measurement, Normalization, and Weights

4.1 OVERVIEW

In decision-making, a commonly occurring problem is how to measure criterion achievement or what evaluation measures to use. It is easy to see why it is extremely important to be able to evaluate alternatives with each criterion. Without useable and appropriate measures, it will be impossible to judge alternatives or preferences. In this chapter, we will discuss how the degree of attainment of criteria can be measured through evaluation measures and their corresponding measurement scales. We also describe different normalization schemes for outcome evaluations. Finally, several techniques for developing criteria importance weights are explained.

4.2 EXAMPLE PROBLEMS

Example problems used to explain different methods of this chapter are described here.

4.2.1 EXAMPLE PROBLEM 4.1

Consider a decision situation involving three alternatives, A_1, A_2, and A_3, and two evaluation criteria, C_1 (benefit-type) is to be maximized and C_2 (cost-type) is to be minimized. The achieved outcome values of the alternatives in each criterion are indicated in Table 4.1.

TABLE 4.1
Problem 4.1 Data

	C_1	C_2
A_1	$v_{11} = 10$	$v_{12} = 0.5$
A_2	$v_{21} = 8$	$v_{22} = -0.2$
A_3	$v_{31} = 12$	$v_{32} = 0.4$

DOI: 10.1201/9781003172291-4

4.2.2 EXAMPLE PROBLEM 4.2

This is a problem that has five criteria and where the decision-maker's (DM's) indicated preferences of criteria are provided through a ranking of the five criteria; a rank of 1 indicates the highest rank and a rank of 5 indicates the lowest rank. Table 4.2 shows the complete rank order indicated by the DM. In this table, C_4 is the most important criterion (rank = 1) and C_3 is the least important criterion (rank = 5).

TABLE 4.2
Problem 4.2 Criteria Rankings by the DM

Criterion	Rank, r_j
C_1	4
C_2	2
C_3	5
C_4	1
C_5	3

4.2.3 EXAMPLE PROBLEM 4.3

Table 4.3 shows an example of the DM-provided pair-wise comparisons where there are five criteria. For example, $c_{13} = 1$ because C_1 is preferred over C_3 and $c_{14} = 0$ because C_1 is not preferred over C_4. The responses are always coded as 1 or 0. The diagonal elements are always 1.

TABLE 4.3
Problem 4.3 Pair-wise Comparisons by the DM

Criterion	C_1	C_2	C_3	C_4	C_5
C_1	1	0	1	0	0
C_2	1	1	1	0	1
C_3	0	0	1	0	0
C_4	1	1	1	1	1
C_5	1	0	1	0	1

4.2.4 EXAMPLE PROBLEM 4.4

There are five criteria, and the following rating scale is used by the DM to rate each criterion. Note that the decision-maker (DM) can use any rating scale deemed appropriate by them.

1 = not important
2 = somewhat important
3 = important
4 = very important
5 = extremely important

Table 4.4 shows the ratings provided by the DM.

TABLE 4.4
Problem 4.4 Criteria Ratings by the DM

Criterion	Rating, r_i
C_1	3
C_2	4
C_3	1
C_4	5
C_5	3

4.2.5 EXAMPLE PROBLEM 4.5

This problem also has five criteria. Table 4.5 shows how a DM has allocated 100 points among five criteria to indicate the DM's criteria preference. Note: the distributed points sum up to 100, the total points allocated for distribution.

TABLE 4.5
Points Allocated by DM to Indicate Preference in Problem 4.5

Criterion	Points Allocated, p_i
C_1	10
C_2	20
C_3	40
C_4	10
C_5	20

4.2.6 EXAMPLE PROBLEM 4.6

Consider that in a specific problem with four criteria, the pair-wise comparison information is provided by a DM as shown in Table 4.6. The DM has used a scale provided by Saaty (1990) for making pair-wise comparisons.

TABLE 4.6
Problem 4.6 Pair-wise Comparisons by the DM

	C_1	C_2	C_3	C_4
C_1	1	1/9	1/3	1/4
C_2	9	1	3	2
C_3	3	1/3	1	1/2
C_4	4	1/2	2	1

If the criterion C_1 is compared to criterion C_2 (considering in this order), the DM can assign a pair-wise comparison value of:

1 to indicate when both C_1 and C_2 are *equally* important
3 to indicate when C_1 is *slightly more* important than C_2
1/3 to indicate when C_1 is *slightly less* important than C_2
5 to indicate when C_1 is *strongly more* important than C_2
1/5 to indicate when C_1 is *strongly less* important than C_2
7 to indicate when C_1 is *very strongly more* important than C_2
1/7 to indicate when C_1 is *very strongly less* important than C_2
9 to indicate when C_1 is *extremely more* important than C_2
1/9 to indicate when C_1 is *extremely less* important than C_2

In such tables, the diagonal elements C_{11}, C_{22}, C_{33}, and C_{44} are always 1. Also, intermediate evaluation numbers 2, 4, 6, and 8 are permitted. Consider row C_3 and column C_1, the value 3 indicates that C_3 is *slightly more important* than C_1; for row C_3 and column C_4, the value 1/2 indicates that C_3 is *in-between equally and slightly less important* than C_4.

4.3   MEASUREMENT AND SCALES OF MEASUREMENT

Criteria measures may be categorized as direct or surrogate (or, proxy). *Direct measures* are those that evaluate the criterion outcomes directly. Such criterion (or outcome) measures and their scales are obvious, or at least seem obvious. For example, a distance measured in Kilometers or temperature measured in Celsius, or cost of raw materials measured in Current Dollars are direct measures. Here, we have a general understanding of the measure and the scale used; and these are in general use in society. Such natural evaluation measures may not, however, be available in all cases or, even if available, may not be practical to use. In situations such as these, the evaluation measure may have to be constructed to use as a surrogate measure. Thus, *surrogate measures* are used in the absence of a usable direct measure. For example, there is no direct measure of passenger comfort in an aircraft seat. However, two surrogate measures, pitch distance between rows of seats and seat width, may be used to measure two dimensions of passenger comfort. Another example is the use of the Dow-Jones (D-J) Industrial Average Index to measure the performance of the stock market. There is no universally accepted or understood measure of stock market performance. As a result, a measure has been developed and D-J Index is the result. After years of use, it has become an accepted measure, though not necessarily correctly understood by the public. Other examples of constructed measures are Grade Point Average (GPA) to measure the academic performance of a student, and dissolved oxygen (DO) concentration in river water as a measure of water quality. In many situations, there may not exist any natural or previously constructed measure. In other cases, the available measures may be too complicated for use. In all such cases, criteria measures may have to be created.

In real-world applications, the distinction between natural and constructed or direct and surrogate measures may not be all that clear or relevant. Important considerations in the selection of criteria measures are that they must be appropriate, usable, and clearly understood by all the participants involved in decision-making.

All evaluation measures involve the construction and use of scales. Measurement scales used in decision-making are usually categorized as ordinal, interval, and ratio. An *ordinal scale* involves the assignment of rank to, the rank ordering of, or categorization of alternatives. Examples of ordinal scaling are when we say alternative A_1 is preferred to alternative A_2 or criterion C_1 is most (or least) important. Only "greater than" (>), "less than" (<), and "equal to" (=) comparisons can be made with an ordinal scale. Technically, no algebraic or other mathematical operations, such as addition, subtraction, etc., can be performed with this scale but we often observe that such mathematical operations are performed (e.g., averages of rank orders to determine group ranking).

An *interval scale* has an arbitrary zero point and a constant unit of measurement. Interval scales can be used to measure differences between measured outcomes. A well-known example of an interval scale is temperatures measured in Celsius. It has an arbitrary zero point (the freezing temperature of water) and a constant unit of measurement where each degree is equal to one-hundredth of the temperature difference between the boiling and freezing points of water. In addition to the comparison operations permitted for ordinal scales, mathematical operations of addition, subtraction, multiplication, and division can be performed with interval scales.

A *ratio scale* has an absolute zero point and a constant unit of measurement. An example of a ratio scale is the distance measured in meters (or yards). Arithmetic operations permitted for interval scales are also permitted for ratio scales. Also, most mathematical transformations (such as multiplications with constants, raising to powers, etc.) are permitted for ratio scales.

4.4 NORMALIZATION

A common problem in decision-making with the use of different units of evaluation measures is that the relative rating of alternatives may change merely because the units of measurement have changed. This can be addressed by normalization. Normalization allows inter-criterion comparison. In the following discussion of normalization, assume that a benefit-type criterion is one in which DM prefers more of it (i.e., more is better) and a cost-type criterion is one in which DM prefers less of it (i.e., less is better). In general, a cost criterion can be transformed mathematically to an equivalent benefit criterion by multiplying by −1 or by taking the inverse of it. We will use Problem 4.1 and its associated Table 4.1 to explain different normalization schemes.

4.4.1 LINEAR NORMALIZATION

Linear normalization converts a measure to a proportion of the way to that between 0 and 1. A measure v_{ij}, outcome for alternative A_i in criterion C_j, is normalized to R_{ij} as follows:

Step 1. Identify the minimum (L) and maximum (H) achievement values for each criterion.

H^*_j = Maximum of v_{ij} for all alternatives for criterion j (for benefit-type criterion)

= Minimum of v_{ij} for all alternatives for criterion j (for all cost-type criterion)

L_{*j} = Minimum of v_{ij} for all alternatives for criterion j (for benefit-type criterion)

= Maximum of v_{ij} for all alternatives for criterion j (for cost type-criterion)

Step 2. Compute the normalized value R_{ij}.

$$R_{ij} = \frac{v_{ij} - L_{*j}}{H^*_j - L_{*j}} \text{ (for benefit-type criterion } j)$$

$$R_{ij} = \frac{L_{*j} - v_{ij}}{L_{*j} - H^*_j} \text{ (for cost-type criterion } j)$$

4.4.1.1 Solution of Problem 4.1 Using Linear Normalization

Step 1. Using Table 4.1 data,

H^*_1 = Maximum of (10, 8, 12) = 12 (a benefit-type criterion)

H^*_2 = Minimum of (0.5, −0.2, 0.4) = −0.2 (a cost-type criterion)

L_{*1} = Minimum of (10, 8, 12) = 8 (a benefit-type criterion)

L_{*2} = Maximum of (0.5, −0.2, 0.4) = 0.5 (a cost-type criterion)

Step 2. The normalized outcome values are indicated as R in Table 4.7. Note that, after normalization, all criteria are transformed to benefit-type criteria (i.e., to be maximized).

TABLE 4.7

Problem 4.1 Outcomes and Their (Linear) Normalized Values

	Benefit-type Criterion C_1		Cost-type Criterion C_2	
	V_1	R_{i1}	V_2	R_{i2}
A_1	10	$R_{11} = \dfrac{v_{11} - L_{*1}}{H^*_1 - L_{*1}} = \dfrac{10 - 8}{12 - 8} = 0.5$	0.5	$R_{12} = \dfrac{L_{*2} - v_{12}}{L_{*2} - H_{*2}} = \dfrac{0.5 - 0.5}{0.5 + 0.2} = 0$
A_2	8	$R_{21} = \dfrac{8 - 8}{12 - 8} = 0$	−0.2	$R_{22} = \dfrac{0.5 + 0.2}{0.5 + 0.2} = 1.0$
A_3	12	$R_{31} = \dfrac{12 - 8}{12 - 8} = 1.0$	0.4	$R_{32} = \dfrac{0.5 - 0.4}{0.5 + 0.2} = 0.14$
H^*_j		$H^*_1 = 12$		$H^*_2 = -0.2$
L_{*j}		$L_{*1} = 8$		$L_{*2} = 0.5$

4.4.2 Vector Normalization

In vector normalization, each criterion outcome v_{ij} is divided by a norm L_{pj} as defined here:

$$R_{ij} = \frac{v_{ij}}{L_{pj}}$$

$$L_{pj} = \left(\sum_{i=1}^{i=n} \left(absolute\ value\ of\ v_{ij} \right)^p \right)^{1/p}$$

where $p = 1$ or 2
$i = (1, 2, \ldots, n)$
$j = (1, 2, \ldots, k)$

With p values of 1 or 2 and the corresponding L_p norms are:

$$L_{1j} = \sum_{i=1}^{i=n} \left(absolute\ value\ of\ v_{ij} \right), \text{p} = 1$$

$$L_{2j} = \left(\sum_{i=1}^{i=n} \left(absolute\ value\ of\ v_{ij} \right)^2 \right)^{1/2}, \text{p} = 2$$

4.4.2.1 Solution of Problem 4.1 Using Vector Normalization

For the alternatives in Table 4.1, the vector-normalized values are computed as shown in Table 4.8 (for $p = 1$) and Table 4.9 (for $p = 2$). In the tables, $|v_{ij}|$ is the absolute value of v_{ij}.

TABLE 4.8
Vector Normalized Values for Problem 4.1. Alternatives with $p = 1$

	Criterion 1 (Benefit-type)		Criterion 2 (Cost-type)	
For p = 1	V_1	R_{i1}	V_2	R_{i2}
A_1	10	$R_{11} = \dfrac{v_{11}}{L_{11}} = \dfrac{10}{30} = 0.33$	0.5	$R_{12} = \dfrac{v_{12}}{L_{12}} = \dfrac{0.5}{1.1} = 0.45$
A_2	8	$R_{21} = \dfrac{8}{30} = 0.27$	-0.2	$R_{22} = \dfrac{-0.2}{1.1} = -0.18$
A_3	12	$R_{31} = \dfrac{12}{30} = 0.4$	0.4	$R_{32} = \dfrac{0.4}{1.1} = 0.36$
L_{1i}		$L_{11} = |10| + |8| + |12| = 30$		$L_{12} = |0.5| + |-0.2| + |0.4| = 1.1$

TABLE 4.9

Vector Normalized Values for Problem 4.1. Alternatives with $p = 2$

	Benefit-type Criterion 1		Cost-type Criterion 2	
For p = 2	V_1	R_{i1}	V_2	R_{i2}
A_1	10	$R_{11} = \dfrac{v_{11}}{L_{11}} = \dfrac{10}{17.55} = 0.57$	0.5	$R_{12} = \dfrac{v_{12}}{L_{12}} = \dfrac{0.5}{0.67} = 0.75$
A_2	8	$R_{21} = \dfrac{8}{17.55} = 0.46$	−0.2	$R_{22} = \dfrac{-0.2}{0.67} = -0.30$
A_3	12	$R_{31} = \dfrac{12}{17.55} = 0.68$	0.4	$R_{32} = \dfrac{0.4}{0.67} = 0.60$
L_{2i}		$L_{21} = \left(\left(|10|\right)^2 + \left(|8|\right)^2 + \left(|12|\right)^2\right)^{1/2}$		$L_{22} = \left(\left(|0.5|\right)^2 + \left(|-0.2|\right)^2 + \left(|0.4|\right)^2\right)^{1/2}$
		$= 17.55$		$= 0.67$

4.5   GENERATION OF WEIGHTS

Many multiple-criteria methods require the use of relative importance weights of criteria. Many of these methods also require ratio scaled weights proportional to the relative value of unit changes in criteria values. Several methods are available for generating these importance weights. These can be grouped into two categories: DM provided and Observer derived. In the DM-provided methods, the DM selects the weights directly, usually through preference or trade-off responses. In Observer-derived methods, the DM indicates the overall evaluation of several alternatives and then weights are formed that are consistent with these evaluations. Regression analysis and linear programming are the commonly used Observer-derived methods. We will only discuss DM-provided methods here.

DM-provided methods require preference judgment from the DM. In extracting numerical judgments about criteria importance, one must be careful about how the judgment questions are framed. It has been suggested that using direct rating (such as in the rating method, rank method, or ratio method) should be preferred to the point allocation methods because the direct rating methods are more reliable.

4.5.1   Weights from Criteria Ranks

This is a simple and commonly used method where only the rank order of the criteria is used for developing the weights.

Step 1. Determine rank order.

DM ranks the criteria in order of decreasing relative importance; that is, the highest-ranked criterion gets a rank of 1. Let r_j represent the rank of the jth criterion.

Step 2. Determine criteria weights.

Determine criterion weight, w_j, as follows:

$$w_j = \frac{k - r_j + 1}{\sum_{i=1}^{i=k} (k - r_i + 1)}, \text{ where } k \text{ is the total number of criteria}$$

This method produces an ordinal scale but may not guarantee the correct type of criterion importance because ranking does not capture the strength of preference information.

4.5.1.1 Solution of Problem 4.2 Using Criteria Ranks Method

Data for Problem 4.2 is used to show how the criteria ranks method can be used to determine criteria weights.

Step 1. The complete rank order of the five criteria is shown in Table 4.2. In this example, DM has indicated that C_4 is the most important criterion and C_3 is the least important criterion, and so on.

Step 2. The last column in Table 4.10 indicates the computed criteria weights.

TABLE 4.10

Computed Criteria Weights from Ranks for Problem 4.2

Criterion	Rank, r_j	Weight, w_j
C_1	4	$w_1 = (5 - 4 + 1)/15 = 0.133$
C_2	2	$w_2 = (5 - 2 + 1)/15 = 0.267$
C_3	5	$w_3 = (5 - 5 + 1)/15 = 0.067$
C_4	1	$w_4 = (5 - 1 + 1)/15 = 0.333$
C_5	3	$w_5 = (5 - 3 + 1)/15 = 0.200$

For example, $w_1 = \dfrac{k - r_j + 1}{\sum_{i=1}^{i=k} (k - r_i + 1)} = \dfrac{5 - 4 + 1}{15} = 0.133$

where, $\sum_{i=1}^{i=k} (k - r_i + 1) = (5 - 4 + 1) + (5 - 2 + 1) + (5 - 5 + 1) +$

$$(5 - 1 + 1) + (5 - 3 + 1) = 15$$

4.5.2 Weights from Pair-wise Ranking of Criteria

When many criteria are considered, it may be easier for the DM to provide a pair-wise ranking instead of a complete ranking. Assuming consistency in pair-wise ranking information, $k(k-1)/2$ such comparisons can be used to derive the complete rank order. The number of times criterion i is ranked higher than all other criteria (in pair-wise comparisons) is used to generate the rank order.

Step 1: Record pair-wise comparison of criteria.

Suppose that for a problem the DM has provided $k(k-1)/2$ pair-wise comparisons of the criteria and the information has been recorded as in Table 4.11. In this table, if c_{ij} indicates the comparison of criterion C_i with criterion C_j, then

$c_{ij} = 1$ if C_i is preferred over C_j
$c_{ij} = 0$ if C_i is not preferred over C_j
Note: c_{ii} is always equal to 1.

TABLE 4.11
A Template for Pair-wise Preference of Criteria Recording

	C_1	C_2	...	C_j	...	C_k	Row Total
C_1	1	c_{12}		c_{1j}		c_{1k}	t_1
C_2	c_{21}	1		c_{2j}		c_{2k}	t_2
.							.
.							.
C_i	c_{i1}	c_{i2}		c_{ij}		c_{ik}	t_i
.							.
.							.
C_k	c_{k1}	c_{k2}		c_{kj}		1	t_k

Step 2: Determine weights.

Find row totals, t_i

$$t_i = \sum_{j=1}^{j=k} c_{ij}$$

Next, find criteria weights as follows:

$$w_i = \frac{t_i}{\sum_{j=1}^{j=k} t_j}$$

4.5.2.1   Solution of Problem 4.3 Using the Pair-wise Ranking of Criteria Method

Step 1. Table 4.3 shows the pair-wise comparisons. For example, $c_{13} = 1$ because C_1 is preferred over C_3 and $c_{14} = 0$ because C_1 is not preferred over C_4.
Step 2. The computed criteria weights are shown in Table 4.12.

TABLE 4.12

Example of Weights from Pair-wise Ranks in Problem 4.3

Criterion	C_1	C_2	C_3	C_4	C_5	t_i = row total	Weight, w_i
C_1	1	0	1	0	0	2	$w_1 = 2/15 = 0.1333$
C_2	1	1	1	0	1	4	$w_2 = 4/15 = 0.2667$
C_3	0	0	1	0	0	1	$w_3 = 1/15 = 0.0667$
C_4	1	1	1	1	1	5	$w_4 = 5/15 = 0.3333$
C_5	1	0	1	0	1	3	$w_5 = 3/15 = 0.2000$

Note: the denominator $15 = 2 + 4 + 1 + 5 + 3$

4.5.3 RATING METHOD

In this method, the DM is asked to rate each criterion judgmentally on an agreed-upon scale (e.g., from 0 to 10). These ratings are then normalized to derive the weights. Even though the method appears to be easy to apply, it fails to assure a ratio scale and it may not even provide the appropriate importance.

Step 1. Select a rating scale.

An appropriate rating scale is agreed to that is clearly understood as to how it is to be used properly.

Step 2. DM rates each criterion.

Using the selected scale, DM provides a rating for each criterion, r_i.

Step 3. *Determine weights.*

Normalize the ratings to determine weights:

$$w_i = \frac{r_i}{\sum_{j=1}^{j=k} r_j}$$

4.5.3.1 Solution of Problem 4.4 Using Rating Method

Step 1. The criteria rating scale used is:
1 = not important
2 = somewhat important
3 = important
4 = very important
5 = extremely important

Steps 2 and 3. Using Table 4.4 data, Table 4.13 shows an example of how the method works. The denominator in the last column is from $(3 + 4 + 1 + 5 + 3) = 16$

TABLE 4.13

Computed Weights from Ratings in Problem 4.4

Criterion	Rating, r_i	Weight, w_i
C_1	3	$w_1 = 3/16 = 0.188$
C_2	4	$w_2 = 4/16 = 0.250$
C_3	1	$w_3 = 1/16 = 0.063$
C_4	5	$w_4 = 5/16 = 0.313$
C_5	3	$w_5 = 3/16 = 0.188$

4.5.4 ALLOCATION METHOD

This method asks the DM to distribute a fixed sum, F, among the criteria considering their relative importance. And then, the normalized values of the distributed scores are the weights. Some studies have suggested that this method can yield a ratio scale and the results can be very close to that of the ratio methods.

Step 1. DM decides on a fixed sum F to be distributed among criteria.

DM decides first on the fixed sum to be distributed (usually, 100 points).

Step 2. Ask DM to distribute the fixed sum amount to each criterion depending on how important each criterion is such that the total of the distributed point equals the agreed fixed sum.

DM allocates that amount F to all the criteria. The more important criteria should be allocated relatively more points. Let p_i represent the points allocated to the ith criterion and F represent the agreed fixed sum. Therefore,

$$\sum_{i=1}^{i=k} p_i = F$$

Step 3. Determine weights.

Weights are determined as follows:

$$w_j = p_j / F$$

4.5.4.1 Solution of Problem 4.5 Using Allocation Method

Step 1. Suppose that the DM decides to use $F = 100$ as the fixed sum to be distributed over the five criteria.

Step 2. Table 4.5 shows the points allocated by the DM among the five criteria.

Step 3. Table 4.14 shows how weights are calculated using the Allocation method.

TABLE 4.14

Computed Weights Using Allocation Method in Problem 4.5

Criterion	Points Allocated, p_i	Weight, w_i
C_1	10	$w_1 = 10/100 = 0.1$
C_2	20	$w_2 = 20/100 = 0.2$
C_3	40	$w_3 = 40/100 = 0.4$
C_4	10	$w_4 = 10/100 = 0.1$
C_5	20	$w_5 = 20/100 = 0.2$

4.5.5 GEOMETRIC MEAN METHOD

A very good approximate method for finding weights from a pair-wise preference table is to find the geometric mean of each row and then normalize those means. A geometric mean is computed by multiplying k numbers and then taking a $1/k$ root of that number.

Step 1. DM provides pair-wise comparisons.

Select a comparison scale. For the k number of criteria, the DM provides $k(k-1)/2$ pair-wise comparisons of the criteria using the preselected scale. A commonly used scale is provided by Saaty (1990) for making a pair-wise comparison of two criteria. For example, if C_1 is compared to C_2 (considering in this order) then the DM would use,

1 to indicate when both C_1 and C_2 are *equally* important
3 to indicate when C_1 is *slightly more* important than C_2
1/3 to indicate when C_1 is *slightly less* important than C_2
5 to indicate when C_1 is *strongly more* important than C_2
1/5 to indicate when C_1 is *strongly less* important than C_2
7 to indicate when C_1 is *very strongly more* important than C_2
1/7 to indicate when C_1 is *very strongly less* important than C_2
9 to indicate when C_1 is *extremely more* important than C_2
1/9 to indicate when C_1 is *extremely less* important than C_2

Step 2. Compute the geometric mean of each row as follows:

Geometric mean of row i, $m_i = \left(\prod_{j=1}^{j=k} a_{ij} \right)^{1/k}$ where, a_{ij} = pair-wise comparison of C_i

with C_j and k is the total number of criteria.

Step 3. Compute weights as follows:

$$w_i = \frac{m_i}{\sum_{j=1}^{j=k} m_j}$$

4.5.5.1 Solution of Problem 4.6 Using Geometric Mean Method

Step 1. Problem 4.6 has four criteria and the pair-wise comparison information provided by the DM is as shown in Table 4.6.

Steps 2 and 3. Computational results of these two steps are shown in Table 4.15.

TABLE 4.15

Problem 4.6 Pair-wise Comparisons by the DM and Computed Weights

	C_1	C_2	C_3	C_4	Geometric Mean, m_i	Weight, w_i
C_1	1	1/9	1/3	1/4	$(1*1/9*1/3*1/4)^{1/4} = 0.310$	$w_1 = 0.31/5.276 = 0.059$
C_2	9	1	3	2	2.711	$w_2 = 0.514$
C_3	3	1/3	1	1/2	0.841	$w_3 = 0.159$
C_4	4	1/2	2	1	1.414	$w_4 = 0.268$
Column Sum					5.276	1.000

5 Single Criterion Decision Problems

5.1 OVERVIEW

Human beings are imperfect information processors. As a result, personal insights about uncertainty and preferences can be limited and misleading. But personal judgments about uncertainty in criterion values are often an important input for decision-making. We know that decision analysis is an iterative process – model building and solution, sensitivity analysis, redefine (goals, alternatives, model, etc.), and redo. In this iterative process, the decision-maker's (DM's) perception of the problem changes, beliefs about the likelihood of uncertain eventualities may change, and preferences may mature as new information becomes available. In this chapter, we explore decision problems that have one evaluation criterion.

5.2 EXAMPLE PROBLEMS

Three example problems are used to describe the solution process of different methods described in this chapter.

5.2.1 EXAMPLE PROBLEM 5.1

Consider a company that has developed five prototypes of a potential new product (A_1, A_2, A_3, A_4, and A_5). It now faces the decision about which of the prototypes to be selected for further development into a marketable new product. The criterion to be used is the estimated dollar sales in the first 2 years after it is introduced in the market. The sales department is reasonably confident about the estimated net sales of the product. Table 5.1 provides the estimated net sales revenue for each prototype. Net-sales revenue = gross sales revenue – production and distribution cost – imputed development and promotion cost.

TABLE 5.1
Estimated Net Sales of the Five Prototypes in 2 Years

Prototype	Estimated Net-sales Revenue, $m
A_1	199
A_2	54
A_3	108
A_4	91
A_5	121

DOI: 10.1201/9781003172291-5

5.2.2 EXAMPLE PROBLEM 5.2

Consider a company that has developed five prototypes of a potential new toy (A_1, A_2, A_3, A_4, and A_5). It now faces the decision about which prototype to be selected for further development into a marketable new product. The future market of this new toy depends on a combination of the economic environment and competitor behavior. The company has no control over what the future economic condition will be and what its competitors' competitive response would be. However, its marketing department has identified four potential states of nature (or scenarios of what the future could be):

S_1 = High growth economy and low competitive response
S_2 = High growth economy but an average competitive response
S_3 = Medium growth economy and average competitive response
S_4 = Low growth economy and low competitive response

The associated pay-off table is shown in Table 5.2. The table values are estimated profit potential on a scale of 1 (extremely low) to 20 (extremely high).

TABLE 5.2
Pay-off Table for Problem 5.2

Alternatives	States of Nature			
	S_1	S_2	S_3	S_4
A_1	17	13	9	5
A_2	18	11	7	4
A_3	17	14	7	3
A_4	20	11	9	2
A_5	15	14	10	4

5.2.3 EXAMPLE PROBLEM 5.3

Consider a company that designs and manufactures indoor security cameras. A current model with a limited algorithmic learning module is being considered for upgrading with an artificial intelligence (AI) infused chip to replace the current algorithmic software. The decision to be made is whether to make the AI-chip upgrade or continue with some software improvements of the current model. The crucial issue is whether the AI-chip will be vastly superior or a mild improvement over the current model with software improvement. If the company decides to develop the new technology, and dependent upon the above two potential uncertain outcomes (vastly superior improvement or mild improvement), the company will face the launch or no-launch decision for the AI-chip. Figure 5.1 shows the decision tree associated with this problem.

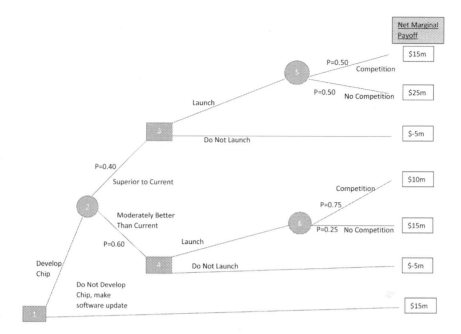

FIGURE 5.1   Decision tree for problem 5.3.

5.3   SINGLE CRITERION DECISION PROBLEMS WITH NO UNCERTAINTY

Some choice situations may not require explicit probabilities. This can happen when the timeframe of the outcome evaluation is relatively short and estimates of the outcome evaluation criteria can be made with a fair degree of certainty. Even when uncertainty exists, there may be a need to make the simplifying assumption of no uncertainty to derive an initial impression about the solution impact and if such simplifying assumption of certainty can be made without much loss of realism.

5.3.1   SINGLE CRITERION OPTIMIZATION METHOD

When there is no uncertainty regarding the outcome values and there is only one evaluation criterion, the solution approach is simple and straightforward, optimize the criterion value.

> *Step 1. Define the problem; identify the alternatives and the evaluation criterion.*

We can use any of the methods described in Chapter 3 for identifying the alternatives and the outcome evaluation criterion to be used (and, whether it is to be maximized or minimized).

Step 2. Compute the evaluation criterion value for each alternative and the preferred alternative is that which optimizes the criterion value.

When the criterion is to be maximized, we choose that alternative that has the maximum evaluation criterion value. In the case of minimization, we choose the alternative with the minimum evaluation value.

5.3.1.1 Solution of Problem 5.1 Using Single Criterion Optimization Method

Step 1. This problem has five alternatives and the criterion to be maximized is the expected net-sales revenue in the first 2 years.

Step 2. Table 5.1 shows the outcome values for the five alternatives. Since the criterion value is to be maximized, preferred alternative is that which maximizes $\{v(A_1), v(A_2), v(A_3), v(A_4), v(A_5)\} = \{199, 54, 108, 91, 121\}$.

The alternative to be chosen is A_1 as it has the maximum criterion value of 199.

5.4 SINGLE CRITERION DECISION PROBLEMS WITH UNCERTAINTY

When there is uncertainty about the future outcome of the decision to be made, we often develop several future outcomes based on potential scenarios of events. But data may not be available on which to base statements of their probabilities of occurrence. Responses in such a case might be to:

a. get additional data or do experiments to develop probabilities (in many situations this may be impossible or would be very costly), or
b. get expert or management judgments to develop probabilities (using, for example, techniques such as the Delphi method), or
c. treat the decision problem using some principles of choice which do not require probabilities as part of input data (i.e., the pay-off matrix does not have probabilities).

Before proceeding further, we will describe 'states of nature' first.

States of Nature: States of nature (sometimes also called Scenarios) in a decision-making situation refer to the future conditions that are likely to occur and over which the decision-maker has no control.

5.4.1 LAPLACE METHOD

In this method, we assume that all the states of nature (S_j) have equal probabilities of occurrence and the criterion is to be maximized. The solution steps are:

Step 1. Define the problem; identify all alternatives; identify possible states of nature; and develop the pay-off matrix.

Once we have defined the problem clearly and completely, we can use any of the methods described in Chapter 3 for identifying the alternatives. Next, the states of nature of the problem are defined as clearly as possible. Finally, we develop the payoff matrix.

Step 2. Normalize the payoff matrix.

Use linear normalization. For H_j and L_j, our suggestion is to use the upper and lower ends of the scale used for developing the payoff matrix (H_j = max of the scale, L_j = min of the scale).

Step 3. Assume equal probabilities for all identified states of nature and determine the expected value (EV) of each alternative as follows:

$$EV\,(A_i) = p_1{*}r_{i1} + p_2{*}r_{i2} + \ldots + p_k{*}r_{ik}, \; i = 1, 2, \ldots, n$$

where A_i is alternative i, r_{ij} is the criterion outcome of A_i for the state of nature S_j, and p_j is the probability that the state of nature will be S_j (p_j = 1/k, and k = total number of states of nature or scenarios).

Step 4. Identify the preferred alternative.

The preferred alternative is that alternative that has the maximum EV.

Preferred A_i = Alternative with the maximum of $\{EV(A_1), EV(A_2), \ldots, EV(A_n)\}$

5.4.1.1 Solution of Problem 5.2 Using Laplace Method

Step 1. The pay-off matrix in Table 5.2 shows the alternatives and the criterion outcome values for each state of nature. The outcomes are total net sales in $m in the first 2 years of product marketing.

Step 2. The scale used for developing the payoff matrix is (20, 1). So, H_j = 20 and L_j = 1. Table 5.3 shows the normalized payoff matrix.

Step 3. Since there are four states of nature, each is assumed to have a 0.25 probability of occurrence (p_j = 1/4). Table 5.3 shows computed results.

As an example of how the EVs are calculated, the EV(A₁) calculation is shown here.

$$EV(A_1) = 0.25{*}0.84 + 0.25{*}0.63 + 0.25{*}0.42 + 0.25{*}0.21 = 0.53$$

TABLE 5.3
Expected Values of Problem 5.2 Alternatives Using Laplace Method

	R_1	R_2	R_3	R_4	
p_j	0.25	0.25	0.25	0.25	EV
A_1	$(17-1)/(20-1) = 0.84$	0.63	0.42	0.21	0.53
A_2	0.89	0.53	0.32	0.16	0.47
A_3	0.84	0.68	0.32	0.11	0.49
A_4	1.00	0.53	0.42	0.05	0.50
A_5	0.74	0.68	0.47	0.16	0.51

Step 4. The preferred alternative is A_1 because $EV(A_1)$ is the maximum of
{0.53, 0.47, 0.49, 0.50, 0.51}.

5.4.2   MAX-MIN METHOD

The Max-Min method assumes that the DM is very pessimistic in outlook and just
wants to choose the best alternative by considering only the worst possible outcomes
(over all the states of nature) of each alternative. Alternatives are characterized by
the minimum criterion achievement amongst its criterion values under various states
of nature. This is a conservative or pessimistic principle that directs attention to the
worst outcomes and tries to make the worst outcome as desirable as possible. The
solution steps are shown here, where each criterion is to be maximized.

*Step 1. Define the problem; identify all alternatives; identify possible states of
nature; and develop the pay-off matrix.*

Once we have defined the problem clearly and completely, we can use any of the
methods described in Chapter 3 for identifying the alternatives. Next, the states of
nature of the problem are defined as clearly as possible. Finally, we develop the pay-
off matrix. Measurement scale boundaries for the criteria values are (H_j = maximum
of scale, L_j = minimum of scale).

*Step 2. Normalize the payoff matrix. Next, identify the minimum criterion
outcome values for each alternative for all states of nature.*

Computation for the normalized value r_{ij} uses the following equation:

$$r_{ij} = (S_{ij} - L_j)/(H_j - L_j) \; i = 1, 2, ..., n \text{ and } j = 1, 2, ..., k$$

$$\text{Row minimum, } M_i = \text{Min of } \{r_{i1}, r_{i2}, ..., r_{ik}\}, i = 1, 2, ..., n$$

Step 3. Identify the preferred alternative.

The preferred alternative is that alternative that has the maximum value of the minimums identified in Step 2.

Preferred A_i = Alternative with the maximum of $\{M_1, M_2, ..., M_n\}$

5.4.2.1 Solution of Problem 5.2 Using Max-Min Method

Step 1. The pay-off matrix in Table 5.2 shows the alternatives and the criterion outcome values for each state of nature. The outcomes are net sales in $m in the first 2 years of product marketing. The scale used for these estimates is (20, 1).

Step 2. We use linear normalization to develop the normalized payoff matrix shown in Table 5.4. The minimum of each alternative for all four states of nature is shown in the last column of Table 5.4. Note that R_1, R_2, R_3, and R_4 columns are copied from Table 5.3.

TABLE 5.4

Minimum Achievements of Alternatives for All States of Nature for Problem 5.2

	R_1	R_2	R_3	R_4	Row Minimum
A_1	0.84	0.63	0.42	0.21	0.21
A_2	0.89	0.53	0.32	0.16	0.16
A_3	0.84	0.68	0.32	0.11	0.11
A_4	1.00	0.53	0.42	0.05	0.05
A_5	0.74	0.68	0.47	0.16	0.16

A calculation example of finding Min (A_1) is shown here:

$$M_1 = \text{Minimum of } \{0.84, 0.63, 0.42, 0.21\} = 0.21$$

Step 3. The preferred alternative has the maximum of the five minimum values.

The preferred alternative is A_1 because it has the maximum of $\{0.21, 0.16, 0.11, 0.05, 0.16\}$.

5.4.3 MIN-MAX REGRET METHOD

In this method, it is assumed that DM wants to minimize the maximum opportunity loss. Opportunity loss is defined as the difference between the maximum value of a criterion and the achieved value of that criterion for an alternative. The solution steps are as shown here, where each criterion is to be maximized.

> *Step 1. Define the problem; identify all alternatives; identify possible states of nature; and develop the pay-off matrix.*

Once we have defined the problem clearly and completely, we can use any of the methods described in Chapter 3 for identifying the alternatives. Next, the states of nature of the problem are defined as clearly as possible. Finally, we develop the payoff matrix.

> *Step 2. Normalize the payoff matrix. Next, identify the maximum values, H_j, for each state of nature (over all alternatives).*

Normalize the payoff matrix as shown in the Max-Min method in Section 5.4.2.

$$H_j = \text{Maximum of } \{r_{1j}, r_{2j}, \ldots, r_{nj}\} \; j = 1, 2, \ldots, k$$

> *Step 3. Develop regret matrix.*

Compute regret matrix entries q_{ij}'s as follows: $q_{ij} = H_j - r_{ij}$, $i = 1, 2, \ldots, n$ and $j = 1, 2, \ldots, k$

Table 5.5 shows the calculations for developing a regret matrix.

TABLE 5.5
Computations for Determining the Regret Matrix

	Q_1	Q_2	...	Q_k
A_1	$q_{11} = H_1 - r_{11}$	$q_{12} = H_2 - r_{12}$	...	$q_{1k} = H_k - r_{1k}$
A_2	$q_{21} = H_1 - r_{21}$	$q_{22} = H_2 - r_{22}$	...	$q_{2k} = H_k - r_{2k}$
...			...	
A_n	$q_{n1} = H_1 - r_{n1}$	$q_{n2} = H_2 - r_{n2}$	...	$q_{nk} = H_k - r_{nk}$

> *Step 4. Identify the maximum regret values for each alternative.*

$$Q^*_i = \text{Maximum of } \{q_{i1}, q_{i2}, \ldots, q_{ik}\}, \; i = 1, 2, \ldots, n$$

> *Step 5. Identify the preferred alternative.*

The preferred alternative is the one associated with minimum regret, minimum Q^*_i.

5.4.3.1 Solution of Problem 5.2 Using Min-Max Regret Method

> *Step 1.* The pay-off matrix in Table 5.2 shows the alternatives and the criterion outcome values for each state of nature. The outcomes are net sales in $m in the first 2 years of product marketing.

Step 2. Table 5.4 has the normalized payoff matrix. Computed results for H_j's are shown here.

$$H_1 = \text{Maximum of } \{0.84, 0.89, 0.84, 1.00, 0.74\} = 1.00$$

$$H_2 = 0.68, \; H_3 = 0.47, \; H_4 = 0.21$$

Step 3. The regret matrix is shown in Table 5.6. Note: the numbers in the table may be different than expected because of rounding up or down.

TABLE 5.6
Regret Matrix for Problem 5.2

	Q_1	Q_2	Q_3	Q_4	Maximum Regret, Q'_i
A_1	$1.00 - 0.84 = 0.16$	0.05	0.05	0.00	0.16
A_2	0.11	$0.68 - 0.53 = 0.16$	0.16	0.05	0.16
A_3	0.16	0.00	0.16	0.11	0.16
A_4	0.00	0.16	0.05	0.16	0.16
A_5	0.26	0.00	0.00	0.05	0.26

Step 4. The maximum regret values for all alternatives are shown in the last column of Table 5.6.

Step 5. The preferred alternative is one of (A_1, A_2, A_3, and A_4) as they all have the same minimum of maximum regret $\{0.16, 0.16, 0.16, 0.16, 0.26\}$.

5.4.4 EXPECTED VALUE METHOD

In the expected value (EV) method, it is presumed that the DM can determine the probability of occurrence for each state of nature (using, for example, the Delphi method).

Step 1. Define the problem; identify all alternatives; identify possible states of nature; and develop the normalized payoff matrix.

Once we have defined the problem clearly and completely, we can use any of the methods described in Chapter 3 for identifying the alternatives. Next, the states of nature of the problem are defined as clearly as possible. Finally, we develop the payoff matrix and its normalized version (using linear normalization).

Step 2. Determine the probability of occurrence of each state of nature.

Methods like the Delphi method can be used to determine probabilities. Let p_j be the probability of the occurrence of the state of nature S_j.

Step 3. Compute the expected value of each alternative.

Compute the expected value of each alternative A_i as follows (the symbol Σ means summation):

$$EV(A_i) = \sum_{j=1}^{k} r_{ij} * p_j$$

Step 4. Select preferred alternative.

The preferred alternative is that alternative with the maximum of $\{EV(A_1), EV(A_2), \ldots, EV(A_n)\}$.

5.4.4.1 Solution of Problem 5.2 Using Expected Value Method

Step 1. The pay-off matrix in Table 5.2 shows the alternatives and the criterion outcome values for each state of nature. The outcomes are net sales in $m in the first 2 years of product marketing. Table 5.7 shows the normalized pay-off matrix. Note that R_1, R_2, R_3, and R_4 columns are copied from Table 5.3.

TABLE 5.7
Expected Values of Alternatives of Problem 5.2

	R_1	R_2	R_3	R_4	
Probability	0.2	0.45	0.25	0.1	EV(A$_i$)
A_1	0.84	0.63	0.42	0.21	0.58
A_2	0.89	0.53	0.32	0.16	0.51
A_3	0.84	0.68	0.32	0.11	0.57
A_4	1.00	0.53	0.42	0.05	0.55
A_5	0.74	0.68	0.47	0.16	0.59

Step 2. Let the determined occurrence probabilities for each state of nature are:

$$p_1 = 0.20, \ p_2 = 0.45, \ p_3 = 0.25, \ p_4 = 0.10$$

note that $p_1 + p_2 + p_3 + p_4$ must equal 1.0.

Step 3. Computed EVs are shown in the last column of Table 5.7.

An example calculation is shown here:

$$EV(A_1) = 0.84*0.20 + 0.63*0.45 + 0.42*0.25 + 0.21*0.10 = 0.58$$

Step 4. The preferred alternative is A_5 because $EV(A_5)$ is the maximum of $\{0.58, 0.51, 0.57, 0.55, 0.59\}$.

5.4.5 DECISION TREE METHOD

When the problem is complex enough to have multiple stages, that is, it involves a sequence of decisions and outcomes, the use of the Decision Tree method is appropriate. It is a tree-like structure similar to the Tree Diagram of Chapter 2, but here it represents a chronological sequence of various decision-outcome developments. The decision tree is built up as a connection of decision nodes and outcome nodes (in sequence). Probabilities are associated with branches from an outcome node. The model will consist of alternating decisions and outcomes, finally terminating in a net pay-off evaluation for each possible path. Solution steps of the Decision Tree method are:

Step 1. Define the problem and collect data (information).

This step involves a clear and complete determination of the problem to be solved including the eventual goal. All necessary data relating to potential decisions as well as the probabilities and pay-offs are collected as the problem would need to be systematically broken down into decision-outcome sequences. We start with the root node that is a goal statement which is the broad goal of the problem that can be systematically broken down into actions. To ease reference to nodes, the nodes will be numbered with the root (goal decision) node being node 1.

Step 2. Starting with the goal decision node, branch off into actions (alternative decisions) from this node.

Add branches for all possible alternatives or outcomes from the goal decision node. A line or arrow represents a branch (which is an alternative or outcome) and connects to an outcome or chance node.

Step 3. Add branches from each chance node.

Branches from each chance node represent possible chance outcomes with associated probabilities of occurrence. The sum of probabilities from all branches from a chance node must add to 1.0.

Step 4. Continue adding decision node-chance node and associated branches till no more decision-chance node pair can be added. The last node should be a chance node and its branches.

At the end of the branches from the last chance node, indicate the net pay-off for that branch.

Step 5. Compute effective pay-offs (EP) at each chance node successively going from the right (the last chance node) to the left till the root decision node is reached. The preferred alternative (decision) is the one with maximum EP.

After the decision tree has been developed, a thorough review of the decision tree should be done to assure that appropriate chronological decision-chance events have been identified along with correct probabilities for branches of chance nodes. We must make sure that the nodes are numbered in chronological order. The EP of a path is computed as shown here for Problem 5.3.

5.4.5.1   Solution of Problem 5.3 Using Decision Tree Method

Step 1. The problem description defines the problem and the decision tree shown in Figure 5.1 shows the collected data.

Steps 2, 3, and 4. See Figure 5.1. The numbers in the right-most column are net marginal (compared to current) cost or pay-off in $m in the first year after introduction.

Step 5. Compute EP for chance nodes 5 and 6,

$$EP(5) = 0.50*\$15m + 0.50*\$25m = \$20m$$

$$EP(6) = 0.75*\$10m + 0.25*\$15m = \$11.25m$$

For decision node 3, we keep EP(5) = $20 m, as it is the maximum of {$20 m, $–5 m} and scratch from further consideration 'Do Not Launch' branch.

For decision node 4, we keep EP(6) = $11.25 m, as it is the maximum of {$11.25 m, $–5 m} and scratch from further consideration 'Do Not Launch' branch.

For chance node 2, EP(2) = 0.40*$20 m + 0.60*$11.25 m = $14.75 m.

Therefore, EP(1) = Maximum of {$14.75 m, $15 m} = $15 m.

The preferred decision is 'Do Not Develop AI chip, Make Software Update' as its EP = $15 m is greater than the EP = $14.75 m of 'Develop AI Chip'. Since the EPs are so close, a review of all estimates should be made to recalculate EP(1).

6 Multiple Criteria Decision Problems

6.1 OVERVIEW

Methods discussed in this chapter involve multiple criteria for evaluating alternate solutions to a decision problem. The evaluation criteria are characterized by their differing measurement scales and their conflicting nature with each other. Because of the complexity of these decision problems, some of the decision methods are much more elaborate and mathematically complex compared to the methods described in previous chapters. Methods in this chapter can be grouped into two categories: (1) those that do not require any preference or trade-off information from the decision-maker (DM) and (2) those that require DM's preference or trade-off information either before solving the problem or during the solution of the problem. For more on many of the methods explained in this chapter, see Hwang and Masud (1979), Hwang and Yoon (1981), and Masud and Ravindran (2008).

6.2 EXAMPLE PROBLEMS

Three example problems will be used to explain the methods in this chapter. These problems are described here.

6.2.1 PROBLEM 6.1

Consider a company trying to choose among five potential subcontractors for supplying a component for a very critical product that has gone through a major design upgrade. The company would like to evaluate these vendors (alternatives) based on three criteria (all to be maximized):

C_1 = vendor's record of maintaining delivery schedule; scale (100 = almost always meets delivery schedule, 50 = usually meets delivery schedule, 10 = frequently fails to meet delivery schedule)

C_2 = vendor's record of staying within initial budget; scale (+10 = often under budget, 0 = usually on budget, −10 = often over budget)

C_3 = vendor's record of meeting quality requirements; scale (10 = superior and usually needs no rework, 5 = average and sometimes needs rework, 0 = extremely poor and often needs rework)

The pay-off matrix in Table 6.1 shows the achievement outcome for each alternative for each criterion. For example, vendor (alternative) A_1 has an outcome of 60 for criterion C_1 (record of maintaining delivery schedule).

DOI: 10.1201/9781003172291-6

TABLE 6.1
Pay-off Matrix for Problem 6.1

	C_1	C_2	C_3
A_1	60	−3	7
A_2	38	8	5
A_3	82	−5	10
A_4	56	2	6
A_5	29	0	9
H_j	82	8	10
L_j	29	−5	5

6.2.2 PROBLEM 6.2

This problem involves levels of a hierarchy of criteria, sub-criteria, and alternatives. This hierarchy will be used to explain the computational steps for the AHP method; Figure 6.1 shows the hierarchy structure of this problem. There are three criteria, C_1, C_2, and C_3, at Level 1; two sub-criteria D_1 and D_2 for criterion C_2 at Level 2; and three alternatives A_1, A_2, and A_3 at Level 3. There are no sub-criteria for C_1 and C_3.

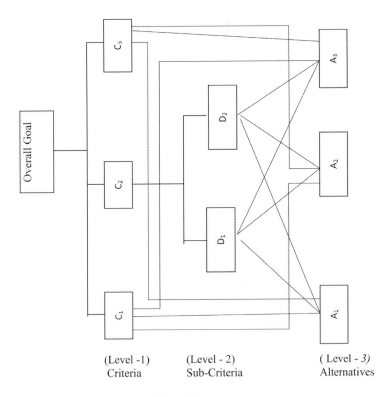

FIGURE 6.1   AHP Hierarchy of Problem 6.2.

At Level 3 the DM has provided pair-wise comparisons of the three alternatives A_1, A_2, and A_3 with respect to C_1, D_1, D_2, and C_3. In Level 2, DM has provided pair-wise comparisons of D_1 and D_2 with respect to C_2. Finally, in Level 1, DM has provided pair-wise comparisons of C_1, C_2, and C_3 with respect to the overall goal. In total, DM has provided six sets of pair-wise comparisons. The pair-wise comparisons or preference judgments can be provided using any appropriate ratio scale. Saaty (1990) has proposed a scale that is used often for providing these pair-wise preferences. In this example problem, the DM has used Saaty's scale which is shown in Table 6.2.

TABLE 6.2
Saaty's Preference Judgment Scale

When Comparing C_i with C_j	Use Value
When C_i is *equally* preferred (or important) as C_j	1
When C_i is *slightly more* preferred (or important) to C_j	3
When C_i is *strongly more* preferred (or important) to C_j	5
When C_i is *very strongly more* preferred (or important) to C_j	7
When C_i is *extremely more* preferred (or important) to C_j	9
To reflect a compromise between scale values above, use	2, 4, 6, 8
When C_i is *slightly less* preferred (or important) to C_j	1/3
When C_i is *strongly less* preferred (or important) to C_j	1/5
When C_i is *very strongly less* preferred (or important) to C_j	1/7
When C_i is *extremely less* preferred (or important) to C_j	1/9

In a pair-wise comparison matrix, the number of rows is always equal to the number of columns and all diagonal elements are 1. All pair-wise comparisons for Problem 6.2 are shown in Tables 6.3, 6.4, and 6.5.

TABLE 6.3
Level 1 Pair-wise Comparisons of Problem 6.2

	C_1	C_2	C_3
C_1	1	1/3	5
C_2	3	1	7
C_3	1/5	1/7	1
Σ	4.20	1.48	13.00

TABLE 6.4
Level 2 Pair-wise Comparisons of Problem 6.2

	D_1	D_2
D_1	1	2
D_2	1/2	1

TABLE 6.5

Level 3 Pair-wise Comparisons of Problem 6.2

	With Respect to C_1		
	A_1	A_2	A_3
A_1	1	2	1
A_2	1/2	1	1
A_3	1	1	1
	With Respect to D_1		
	A_1	A_2	A_3
A_1	1	2	3
A_2	1/2	1	2
A_3	1/3	1/2	1
	With Respect to D_2		
	A_1	A_2	A_3
A_1	1	2	1/7
A_2	1/2	1	1/9
A_3	7	9	1
	With Respect to C_3		
	A_1	A_2	A_3
A_1	1	2	4
A_2	1/2	1	3
A_3	1/4	1/3	1

6.2.3 PROBLEM 6.3

This is also a problem that involves levels of hierarchy like Problem 6.2 but has a much simpler structure as shown in Figure 6.2. This problem involves three criteria in Level 1 and five alternatives in Level 2. Table 6.6 shows the pair-wise judgments, v_{ij}, provided by DM for Level 1 of the hierarchy.

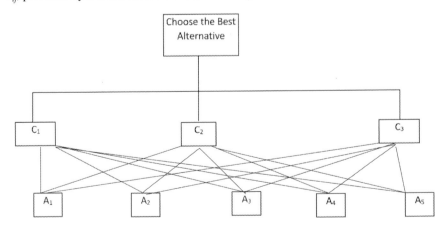

FIGURE 6.2   AHP hierarchy of problem 6.3.

TABLE 6.6
Level 1 Pair-wise Comparisons for Problem 6.3

	C_1	C_2	C_3
C_1	1	1/3	5
C_2	3	1	7
C_3	1/3	1/7	1

For Level 2, the DM does not need to provide pair-wise comparisons as those can be derived from the ratios of the measured outcomes shown in the pay-off table in Table 6.1. All criteria are to be maximized.

6.3 MULTIPLE CRITERIA METHODS REQUIRING NO PREFERENCE INFORMATION

Methods discussed in this section are characterized by what they do not require, preference information from the DM. In these methods, the preferred solution is obtained instead by making some intuitively appealing but somewhat simple preference assumptions, all without any input from the DM.

6.3.1 MIN-MAX REGRET METHOD

In this method, it is assumed that the DM wants to minimize the maximum opportunity loss. Opportunity loss is defined as the difference between the ideal solution value of a criterion and the achieved value of that criterion in an alternative. Geometrically, this method finds a solution that minimizes the maximum normalized distance from the ideal solution along each criterion value for all available alternatives. The solution steps of this method are given here.

Step 1. Define the problem and develop the associated pay-off matrix.

Define and describe the decision problem as clearly as possible, including the potential alternatives, the decision criteria, and the measurement scales of outcomes for each criterion. Next, collect outcome achievement data for each alternative i for each criterion j, v_{ij}. Construct a pay-off table like Table 6.7 where rows are the alternatives,

TABLE 6.7
Pay-off Table for a Decision Problem

	C_1	C_2	...	C_k
A_1	v_{11}	v_{12}	...	v_{1k}
A_2	v_{21}	v_{22}	...	v_{2k}
...			...	
A_n	v_{n1}	v_{n2}	...	v_{nk}
H^*	$H^*_1 = \text{Max}(v_{11}, v_{21}, ..., v_{n1})$	H^*_2	...	H^*_k
L_*	$L_{*1} = \text{Min}(v_{11}, v_{21}, ..., v_{n1})$	L_{*2}	...	L_{*k}

columns are the criteria, and cell values are the v_{ij}. Note: all criteria are to be maximized (or converted to be maximized).

Step 2. Identify ideal and anti-ideal solutions.

The last two rows in Table 6.7 show how to find ideal solution, H*, and anti-ideal solution, L*.

Step 3. Develop normalized regret matrix, Q.

Compute regret matrix entries q_{ij}'s as follows:

$$q_{ij} = \frac{H_j^* - v_{ij}}{H_j^* - L_{*j}}, \; i = 1, 2, ..., n \text{ and } j = 1, 2, ..., k$$

Table 6.8 shows the calculations involved in creating a normalized regret matrix.

TABLE 6.8
Normalized Regret Matrix

	Q_1	Q_2	...	Q_k
A_1	$q_{11} = \dfrac{H_1^* - v_{11}}{H_1^* - L_{*1}}$	$q_{12} = \dfrac{H_2^* - v_{12}}{H_2^* - L_{*2}}$	...	q_{1k}
A_2	q_{21}	$q_{22} = \dfrac{H_2^* - v_{22}}{H_2^* - L_{*2}}$	...	q_{2k}
...			...	
A_n	q_{n1}	q_{n2}	...	q_{nk}

Step 4. Identify the maximum regret values for all alternatives.

Table 6.9 shows how to identify maximum regrets for an alternative.

TABLE 6.9
Maximum Regrets for All Alternatives

	Q^*
A_1	$Q^*_1 = \text{Max} \{q_{11}, q_{12}, ..., q_{1k}\}$
A_2	$Q^*_2 = \text{Max} \{q_{21}, q_{22}, ..., q_{2k}\}$
...	...
A_n	$Q^*_n = \text{Max} \{q_{n1}, q_{n2}, ..., q_{nk}\}$

Step 5. Identify the preferred solution.

Determine $Q_{min} = \{Min\ (Q^*_1, Q^*_2, \ldots, Q^*_n\}$. The preferred alternative is the alternative associated with Q_{min}.

6.3.1.1 Solution of Problem 6.1 Using Min-Max Regret Method

Step 1. Table 6.1 shows the problem pay-off matrix.
Step 2. The last two rows of Table 6.1 show the ideal solution, H^*, and the anti-ideal solution, L_*.
Step 3. Table 6.10 shows the calculated normalized regret matrix.

TABLE 6.10
Normalized Regret Matrix of Problem 6.1 for the Min-Max Regret Method

	Q_1	Q_2	Q_3
A_1	$\dfrac{82-60}{82-29}=0.415$	0.846	$\dfrac{10-7}{10-5}=0.60$
A_2	0.830	0.0	1.0
A_3	0.0	1.0	0.0
A_4	$\dfrac{82-56}{82-29}=0.491$	0.462	0.80
A_5	1.0	0.615	0.20

Step 4. The maximum regret values for the alternatives are calculated as follows:

$$Q^*_1 = \text{Maximum of } (0.415, 0.846, 0.60) = 0.846$$
$$Q^*_2 = 1.0$$
$$Q^*_3 = 1.0$$
$$Q^*_4 = 0.80$$
$$Q^*_5 = 1.0$$

Step 5. To identify the preferred alternative, identify the alternative that has the minimum of maximum regret as follows:

$$Q_{min} = \text{Minimum of } (0.846, 1.0, 1.0, 0.80, 1.0) = 0.80.$$

Therefore, the preferred alternative is A_4 as it has the minimum Q^*_i value of 0.80.

6.3.2 COMPROMISE PROGRAMMING METHOD

Compromise Programming (CP) method identifies the preferred alternative that is as close to the ideal solution as possible. That is, it identifies the solution whose distance from the ideal solution is minimum. The CP method is also known as the Global Criterion method. In CP, the distance is measured with one of the metrics, M_p, as defined here. Distance is normalized first to make it comparable across criteria units. We will use linear normalization.

Step 1. Define the problem and develop the associated pay-off matrix.

Define and describe the decision problem as clearly as possible, including the potential alternatives, the decision criteria, and the measurement scales of outcomes for each criterion. Next, collect outcome achievement data for each alternative i for each criterion j, v_{ij}. Construct a pay-off table like Table 6.7, where rows are the alternatives, columns are the criteria, and cell values are the v_{ij}.

Step 2. Identify ideal and anti-ideal solutions.

The last two rows in Table 6.7 show how to find ideal solution, H*, and anti-ideal solution, L*.

Step 3. Choose the metric, M_p, to be used. Compute the chosen metric value.

Commonly used metrics are $M_{p=1}$, $M_{p=2}$, and $M_{p=\infty}$; where ∞ represents infinity. Only one of the metrics M_1 or M_2 or M_∞ should be used. However, for the sake of explaining all aspects of the method, we will show in this section computations for all three metrics. Compute the metric values for the decision alternatives (based on the p chosen in Step 3). Tables 6.11, 6.12, and 6.13 show the computational details for each metric.

TABLE 6.11

CP Metric Values Computation with p = 1

For $p = 1$	$M_{1i} = \sum_{j=1}^{j=k} \dfrac{H_j^* - v_{ij}}{H_j^* - L_{j*}}$
A_1	$M_{11} = \dfrac{H_1^* - v_{11}}{H_1^* - L_{1*}} + \dfrac{H_2^* - v_{12}}{H_2^* - L_{2*}} + \ldots + \dfrac{H_k^* - v_{1k}}{H_k^* - L_{k*}}$
A_2	$M_{12} = \dfrac{H_1^* - v_{21}}{H_1^* - L_{1*}} + \dfrac{H_2^* - v_{22}}{H_2^* - L_{2*}} + \ldots + \dfrac{H_k^* - v_{2k}}{H_k^* - L_{k*}}$
...	...
A_n	$M_{1n} = \dfrac{H_1^* - v_{n1}}{H_1^* - L_{1*}} + \dfrac{H_2^* - v_{n2}}{H_2^* - L_{2*}} + \ldots + \dfrac{H_k^* - v_{nk}}{H_k^* - L_{k*}}$

TABLE 6.12
CP Metric Values Computation with p = 2

For p = 2
$$M_{2i} = \left(\sum_{j=1}^{j=k} \left(\frac{H_j^* - v_{ij}}{H_j^* - L_{j*}} \right)^2 \right)^{1/2}$$

A_1
$$M_{21} = \left(\left(\frac{H_1^* - v_{11}}{H_1^* - L_{1*}} \right)^2 + \left(\frac{H_2^* - v_{12}}{H_2^* - L_{2*}} \right)^2 + ... + \left(\frac{H_k^* - v_{1k}}{H_k^* - L_{k*}} \right)^2 \right)^{1/2}$$

A_2
$$M_{22} = \left(\left(\frac{H_1^* - v_{21}}{H_1^* - L_{1*}} \right)^2 + \left(\frac{H_2^* - v_{22}}{H_2^* - L_{2*}} \right)^2 + ... + \left(\frac{H_k^* - v_{2k}}{H_k^* - L_{k*}} \right)^2 \right)^{1/2}$$

...

A_n
$$M_{2n} = \left(\left(\frac{H_1^* - v_{n1}}{H_1^* - L_{1*}} \right)^2 + \left(\frac{H_2^* - v_{n2}}{H_2^* - L_{2*}} \right)^2 + ... + \left(\frac{H_k^* - v_{nk}}{H_k^* - L_{k*}} \right)^2 \right)^{1/2}$$

TABLE 6.13
CP Metric Values Computation with p = ∞

For p = ∞
$$M_{\infty j} = \text{Max} \left\{ \frac{H_1^* - v_{i1}}{H_1^* - L_{*1}}, \frac{H_2^* - v_{i2}}{H_2^* - L_{2*}}, ..., \frac{H_k^* - v_{ik}}{H_k^* - L_{k*}} \right\}$$

A_1
$$M_{\infty 1} = \text{Max} \left\{ \frac{H_1^* - v_{11}}{H_1^* - L_{1*}}, \frac{H_2^* - v_{12}}{H_2^* - L_{2*}}, ..., \frac{H_k^* - v_{1k}}{H_k^* - L_{k*}} \right\}$$

A_2
$$M_{\infty 2} = \text{Max} \left\{ \frac{H_1^* - v_{21}}{H_1^* - L_{1*}}, \frac{H_2^* - v_{22}}{H_2^* - L_{2*}}, ..., \frac{H_k^* - v_{2k}}{H_k^* - L_{k*}} \right\}$$

...

A_n
$$M_{\infty n} = \text{Max} \left\{ \frac{H_1^* - v_{n1}}{H_1^* - L_{1*}}, \frac{H_2^* - v_{n2}}{H_2^* - L_{2*}}, ..., \frac{H_k^* - v_{nk}}{H_k^* - L_{k*}} \right\}$$

Step 4. Identify the preferred solution.

The preferred solution is the alternative that has the minimum M_p value.

6.3.2.1 Solution of Problem 6.1 Using Compromise Programming Method

Step 1. Table 6.1 shows the problem pay-off matrix.
Step 2. The last two rows of Table 6.1 show the ideal solution, H*, and the anti-ideal solution, L*.
Step 3. We should choose only one of p = 1 or p = 2 or p = ∞. But, for demonstration purposes, calculations with all three metrics will be shown. Computed metric values are shown in Tables 6.14 (for p = 1), 6.15 (for p = 2), and 6.16 (for p = ∞).

TABLE 6.14
Problem 6.1 Metric Values for p = 1

For p = 1 $M_{1i} = \dfrac{82 - v_{i1}}{82 - 29} + \dfrac{8 - v_{i2}}{8 - (-5)} + \dfrac{10 - v_{i3}}{10 - 5}$

A_1 $M_{11} = \dfrac{82 - 60}{82 - 29} + \dfrac{8 - (-3)}{8 - (-5)} + \dfrac{10 - 7}{10 - 5} = 1.86$

A_2 $M_{12} = \dfrac{82 - 38}{82 - 29} + \dfrac{8 - 8}{8 - (-5)} + \dfrac{10 - 5}{10 - 5} = 1.83$

A_3 $M_{13} = \dfrac{82 - 82}{82 - 29} + \dfrac{29 - (-5)}{5 - (-5)} + \dfrac{10 - 10}{10 - 5} = 1.0*$

A_4 $M_{14} = \dfrac{82 - 56}{82 - 29} + \dfrac{8 - 2}{8 - (-5)} + \dfrac{10 - 6}{10 - 5} = 1.75$

A_5 $M_{15} = \dfrac{82 - 29}{82 - 29} + \dfrac{8 - 0}{8 - (-5)} + \dfrac{10 - 9}{10 - 5} = 1.82$

TABLE 6.15
Problem 6.1 Metric Values for p = 2

For p = 2 $M_{2i} = \left[\left(\dfrac{82 - v_{i1}}{82 - 29} \right)^2 + \left(\dfrac{8 - v_{i2}}{8 - (-5)} \right)^2 + \left(\dfrac{10 - v_{i3}}{10 - 5} \right)^2 \right]^{1/2}$

A_1 $M_{21} = \left[\left(\dfrac{82 - 60}{82 - 29} \right)^2 + \left(\dfrac{8 - (-3)}{8 - (-5)} \right)^2 + \left(\dfrac{10 - 7}{10 - 5} \right)^2 \right]^{1/2} = 1.12$

A_2 $M_{22} = \left[\left(\dfrac{82 - 38}{82 - 29} \right)^2 + \left(\dfrac{8 - 8}{8 - (-5)} \right)^2 + \left(\dfrac{10 - 5}{10 - 5} \right)^2 \right]^{1/2} = 1.30$

A_3 $M_{23} = \left[\left(\dfrac{82 - 82}{82 - 29} \right)^2 + \left(\dfrac{8 - (-5)}{8 - (-5)} \right)^2 + \left(\dfrac{10 - 10}{10 - 5} \right)^2 \right]^{1/2} = 1.0*$

A_4 $M_{24} = \left[\left(\dfrac{82 - 56}{82 - 29} \right)^2 + \left(\dfrac{8 - 2}{8 - (-5)} \right)^2 + \left(\dfrac{10 - 6}{10 - 5} \right)^2 \right]^{1/2} = 1.05$

A_5 $M_{25} = \left[\left(\dfrac{82 - 29}{82 - 29} \right)^2 + \left(\dfrac{8 - 0}{8 - (-5)} \right)^2 + \left(\dfrac{10 - 9}{10 - 5} \right)^2 \right]^{1/2} = 1.19$

TABLE 6.16
Problem 6.1 Metric Values for $p = \infty$

For $p = \infty$ $M_{\infty j} = \text{Max} \left\{ \left(\dfrac{82 - v_{i1}}{82 - 29} \right), \left(\dfrac{8 - v_{i2}}{8 - (-5)} \right), \left(\dfrac{10 - v_{i3}}{10 - 5} \right) \right\}$

A_1 $M_{\infty 1} = \text{Max} \left\{ \left(\dfrac{82 - 60}{82 - 29} \right), \left(\dfrac{8 - (-3)}{8 - (-5)} \right), \left(\dfrac{10 - 7}{10 - 5} \right) \right\} = \text{Max} \{0.415, 0.846, 0.6\} = 0.846$

A_2 $M_{\infty 2} = \text{Max} \left\{ \left(\dfrac{82 - 38}{82 - 29} \right), \left(\dfrac{8 - 8}{8 - (-5)} \right), \left(\dfrac{10 - 5}{10 - 5} \right) \right\} = \text{Max} \{0.83, 0, 1.0\} = 1.0$

A_3 $M_{\infty 3} = \text{Max} \left\{ \left(\dfrac{82 - 82}{82 - 29} \right), \left(\dfrac{8 - (-5)}{8 - (-5)} \right), \left(\dfrac{10 - 10}{10 - 5} \right) \right\} = \text{Max} \{0, 1.0, 0\} = 1.0$

A_4 $M_{\infty 4} = \text{Max} \left\{ \left(\dfrac{82 - 56}{82 - 29} \right), \left(\dfrac{8 - 2}{8 - (-5)} \right), \left(\dfrac{10 - 6}{10 - 5} \right) \right\} = \text{Max} \{0.491, 0.462, 0.80\} = 0.80^*$

A_5 $M_{\infty 5} = \text{Max} \left\{ \left(\dfrac{82 - 29}{82 - 29} \right), \left(\dfrac{8 - 0}{8 - (-5)} \right), \left(\dfrac{10 - 9}{10 - 5} \right) \right\} = \text{Max} \{1.0, 0.615, 0.20\} = 1.0$

Step 4. Preferred solution.
   If $p = 1$ is used then minimum of (1.86, 1.83, 1.0, 1.75, 1.82) = 1.0; select A_3
   If $p = 2$ is used then minimum of (1.25, 1.69, 1.0, 1.09, 1.42) = 1.0; select A_3
   If $p = \infty$ is used then minimum of (0.846, 1.0, 1.0, 0.80, 1.0) = 0.80; select A_4

6.4   MULTIPLE CRITERIA METHODS REQUIRING PREFERENCE INFORMATION

Methods that require DM's preference information are:

- A Priori Preference methods that rely on receiving some preference judgment from the DM before solving the decision problem. The type of information needed vary from method to method.
- Interactive methods are those that incorporate the DM into the solution process. In these methods, the DM is asked to provide some preference or trade-off information during the solution process. The DM is usually asked to provide preference for one or more solutions or a trade-off preference around a current solution. That information is then used to develop an improved solution, which becomes the current solution. This iterative process continues until the stopping conditions are satisfied. By incorporating the DM into the solution process, the chance of acceptance of the final solution by the DM is increased significantly. The iterative request for information also provides learning for DM about the nature of the solution space.

However, these come at a cost of increased information and time demand from the DM, which the DM might not be willing to provide.

Methods described in this chapter fall in the category of a priori preference methods.

6.4.1 EFFECTIVE VALUE METHOD

In this method, it is assumed that the DM is trying to maximize the overall value of the selected alternative. The trade-off preference, in the form of criteria weights, is determined before the problem is solved.

Step 1. Define the problem, develop the pay-off matrix, and identify the ideal and anti-ideal solutions.

Define and describe the decision problem as clearly as possible, including the potential alternatives, the decision criteria, and the measurement scales of outcomes for each criterion. Next, collect outcome achievement data for each alternative i for each criterion j, v_{ij}. Construct a pay-off table like Table 6.7, where rows are the alternatives, columns are the criteria, and cell values are the v_{ij}. The last two rows in Table 6.7 show how to find ideal solution, H^*, and anti-ideal solution, L_*.

Step 2. Develop normalized pay-off matrix.

The normalized pay-off matrix, using a linear normalization, is calculated as shown in Table 6.17.

TABLE 6.17
Normalized Pay-off Matrix, R, for the Effective Value Method

	R_1	R_2	...	R_k
A_1	$r_{11} = \dfrac{v_{11} - L_1}{H_1 - L_1}$	$r_{12} = \dfrac{v_{12} - L_2}{H_2 - L_2}$	...	r_{1k}
A_2	r_{21}	$r_{22} = \dfrac{v_{22} - L_2}{H_2 - L_2}$	...	r_{2k}
...			...	
A_n	r_{n1}	r_{n2}	...	r_{nk}

Step 3. Determine criteria weights.

Using any of the methods described in Chapter 4, the DM provides preference information to determine the criteria weights, w_j. Note that each $w_j > 0$ and the weights must sum up to 1.0.

Step 4. Compute the weighted effective value for all alternatives.

$$EV(A_i) = w_1 * r_{i1} + w_2 * r_{i2} + \ldots + w_k * r_{ik}, i = 1, 2, \ldots, n$$

Step 5. Select as the preferred alternative the one with the maximum weighted effective value.

The preferred alternative is the one with the maximum of $\{EV(A_1), EV(A_2), \ldots, EV(A_n)\}$.

6.4.1.1   Solution of Problem 6.1 Using Effective Value Method

Step 1. Table 6.1 shows the problem pay-off matrix.
Step 2. Table 6.18 is the normalized pay-off matrix.

TABLE 6.18
Normalized Pay-off Matrix of Problem 6.1 Using the Effective Value Method

	R_1	R_2	R_3
A_1	$\dfrac{60-29}{82-29}=0.585$	$\dfrac{(-3)-(-5)}{8-(-5)}=0.154$	$\dfrac{7-5}{10-5}=0.40$
A_2	$\dfrac{38-29}{82-29}=0.170$	$\dfrac{8-(-5)}{8-(-5)}=1.0$	$\dfrac{5-5}{10-5}=0$
A_3	$\dfrac{82-29}{82-29}=1.0$	$\dfrac{(-5)-(-5)}{8-(-5)}=0$	$\dfrac{10-5}{10-5}=1.0$
A_4	$\dfrac{56-29}{82-29}=0.509$	$\dfrac{2-(-5)}{8-(-5)}=0.538$	$\dfrac{6-5}{10-5}=0.20$
A_5	$\dfrac{29-29}{82-29}=0$	$\dfrac{0-(-5)}{8-(-5)}=0.385$	$\dfrac{9-5}{10-5}=0.80$

Step 3. With DM's participation, the criteria weights have been determined as:

$$W = \left(w_1 = 0.25,\ w_2 = 0.45,\ w_3 = 0.30\right)$$

Step 4.
   $EV(A_1) = 0.25*0.585 + 0.45*0.154 + 0.30*0.40 = 0.336$
   $EV(A_2) = 0.25*0.170 + 0.45*1.0 + 0.30*0 = 0.493$
   $EV(A_3) = 0.25*1.0 + 0.45*0 + 0.30*1.0 = 0.550*$
   $EV(A_4) = 0.25*0.509 + 0.45*0.538 + 0.30*0.20 = 0.429$
   $EV(A_5) = 0.25*0 + 0.45*0.385 + 0.30*0.80 = 0.413$
Step 5. The preferred alternative is A_3 as it has the maximum $EV(A_1)$ value.

6.4.2   TOPSIS Method

TOPSIS (technique for order preference by similarity to ideal solution) was originally proposed by Hwang and Yoon (1981). TOPSIS operates on the principle that the preferred alternative should simultaneously be closest to the ideal solution, H^*, and farthest from the anti-ideal solution, L_*. The method uses an index, which combines the closeness of an alternative to the ideal solution with its remoteness from the anti-ideal solution. The alternative that maximizes this index value is the preferred alternative. Also, the alternatives can be rank-ordered based on the index value. The steps of this method are shown here.

Step 1. Define the problem and determine the relative importance weights of each criterion.

Define and describe the decision problem as clearly as possible, including the potential alternatives, the decision criteria, and the measurement scales of outcomes for each criterion. Using any of the methods described in Chapter 4, the DM provides preference information to determine the criteria weights, w_j. Note that each $w_j > 0$ and the weights must sum up to 1.0.

Step 2. Determine normalized pay-off matrix, R.

In TOPSIS, the pay-off matrix is normalized as follows:

$$r_{ij} = \frac{v_{ij}}{\left(\sum_{i=1}^{i=n} v_{ij}^2\right)^{1/2}}, \ i = 1, 2, \ldots, n \ and \ j = 1, 2, \ldots, k$$

Step 3. Compute weighted pay-off matrix, Q.

$$q_{ij} = w_j * r_{ij}, \ i = 1, 2, \ldots, n \ and \ j = 1, 2, \ldots, k$$

Step 4. Identify ideal solution, H, and anti-ideal solution, L.

$$H_j = Maximum \ of\left(q_{1j}, q_{2j}, \ldots, q_{nj}\right), \ j = 1, 2, \ldots, k$$

$$L_j = Minimum \ of\left(q_{1j}, q_{2j}, \ldots, q_{nj}\right), \ j = 1, 2, \ldots, k$$

Step 5. Compute separation measures P^* and P_*.

$$P_i^* = \left[\sum_{j=1}^{j=k}\left(q_{ij} - H_j\right)^2\right]^{1/2}, \ i = 1, 2, \ldots, n$$

$$P_{*i} = \left[\sum_{j=1}^{j=k}\left(q_{ij} - L_j\right)^2\right]^{1/2}, \ i = 1, 2, \ldots, n$$

where, P_i^* is the distance of A_i from the ideal solution and P_{*i} is the distance of A_i from the negative-ideal solution.

Step 6. Compute the similarity index, D_i.

$$D_i = \frac{P_{*i}}{\left(P_i^* + P_{*i}\right)}, \ i = 1, 2, \ldots, n$$

Note that $0 \leq D_i \leq 1$ and $D_i = 0$ when A_i is the negative-ideal solution and $D_i = 1$ when A_i is the ideal solution.

Step 7. Identify the preferred alternative.

The preferred alternative is that alternative that has the maximum D_i value. All the alternatives can be rank-ordered by the D_i values, maximum to minimum.

6.4.2.1 Solution of Problem 6.1 Using TOPSIS Method

Step 1. The problem definition is provided by the problem statement of Problem 6.1. Suppose that using Chapter 4 methods, the following criterion weights are determined:

$$W = (w_1 = 0.25, w_2 = 0.45, w_3 = 0.30)$$

Step 2. Using Table 6.1 pay-off matrix, the normalized pay-off matrix, R, is determined as shown in Table 6.19.

TABLE 6.19
Normalized Pay-off Matrix for Problem 6.1 Using TOPSIS

	R_1	R_2	R_3
A_1	$(60/125.48) = 0.478$	-0.297	0.410
A_2	0.303	0.792	0.293
A_3	$(82/125.48) = 0.653$	-0.495	0.586
A_4	0.446	0.198	0.352
A_5	0.231	0	0.528
$\left(\sum_{i=1}^{i=n} v_{ij}^2\right)^{1/2}$	$(60^2 + 38^2 + 82^2 + 56^2 + 29^2)^{1/2} = 125.48$	10.10	17.06

Step 3. Weighted pay-off matrix, Q, is shown in Table 6.20.

TABLE 6.20
Weighted Pay-off Matrix for Problem 6.1 Using TOPSIS

	Q_1	Q_2	Q_3
A_1	$0.25*0.478 = 0.120$	$0.45*(-0.297) = -0.134$	$0.30*0.410 = 0.123$
A_2	0.076	0.356	0.088
A_3	0.163	-0.223	0.176
A_4	0.112	0.089	0.106
A_5	0.058	0	0.158
w_j	0.25	0.45	0.30
H_j	0.163	0.356	0.176
L_j	0.058	-0.223	0.088

Step 4. Ideal and anti-ideal solutions are shown in the last two rows of Table 6.20.

Step 5. Separation measures P^* and P_* are computed as follows:

$P_1^* = [(q_{11} - H_1)^2 + (q_{12} - H_2)^2 + (q_{13} - H_3)^2]^{1/2} = [(0.120 - 0.163)^2 + (-0.134 - 0.356)^2 + (0.123 - 0.176)^2]^{1/2} = 0.494$

$P_{*1} = [(q_{11} - L_1)^2 + (q_{12} - L_2)^2 + (q_{13} - L_3)^2]^{1/2} = [(0.120 - 0.058)^2 + (-0.134 - (-0.223))^2 + (0.123 - 0.088)^2]^{1/2} = 0.114$

$P_2^* = [(q_{21} - H_1)^2 + (q_{22} - H_2)^2 + (q_{23} - H_3)^2]^{1/2} = [(0.076 - 0.163)^2 + (0.356 - 0.356)^2 + (0.088 - 0.176)^2]^{1/2} = 0.124$

$P_{*2} = [(q_{21} - L_1)^2 + (q_{22} - L_2)^2 + (q_{23} - L_3)^2]^{1/2} = [(0.076 - 0.058)^2 + (0.356 - (-0.223))^2 + (0.088 - 0.088)^2]^{1/2} = 0.580$

$P_3^* = 0.579 \quad P_{*3} = 0.137$

$P_4^* = 0.281 \quad P_{*4} = 0.317$

$P_5^* = 0.372 \quad P_{*5} = 0.234$

Step 6. Computed similarity index, D, is shown below:

$D_1^* = P_{1*}/(P_1^* + P_{1*}) = 0.114/(0.495 + 0.114) = 0.187$

$D_2^* = 0.580/(0.124 + 0.580) = 0.824*$

$D_3^* = 0.137/(0.579 + 0.137) = 0.191$

$D_4^* = 0.317/(0.281 + 0.317) = 0.530$

$D_5^* = 0.234/(0.372 + 0.234) = 0.386$

Step 7. Rank order of all five alternatives is:

(most preferred to least preferred) $= A_2, A_4, A_5, A_3, A_1$

6.4.3 ANALYTIC HIERARCHY PROCESS (AHP) METHOD

AHP method was first proposed by Saaty in the 1970s. We describe below a modified-solution approach proposed by Golden and Wang (1989). In AHP, the decision problem is first structured in levels of a hierarchy. At the top level is the goal or overall purpose of the problem. The subsequent levels represent criteria, sub-criteria, etc. The last level always represents the decision alternatives. The procedure for determining the composite evaluation of alternatives is based on maximizing the 'overall goal' at the top of the hierarchy (i.e., Level 0). The AHP method is computationally the most involved and complex of all the methods described in this book. For any reasonable size problem, it may become unwieldy even if using a computer-based spreadsheet for all computations. In such a case, we may need to use an available specialized software implementing the AHP method. Several such specialized software can be found with a simple web search. A note of caution: the computational results from these specialized softwares may be slightly different from the method explained here because those usually implement the Saaty proposed approach instead of the simplified approach suggested by Golden and Wang that is explained here.

Step 1. Define the problem and develop the hierarchy structure of the decision problem.

Define and describe the decision problem as clearly as possible, including the potential alternatives, the decision criteria, sub-criteria associated with any of the criterion, and the measurement scales for criteria that are directly measurable. Figure 6.3

shows a typical hierarchical problem representation for AHP. At the top (or left if horizontally drawn) is Level 0 that shows the overall goal. Next, at Level 1, the main criteria; next, at Level 2 the sub-criteria (if any or some main criteria can be expanded to their sub-criteria). This will continue till the last level, Level l, where only the alternatives are listed. We suggest that to keep the structure easier to follow and relatively less complicated, do not add any level after Level 3 (where the alternatives would be indicated).

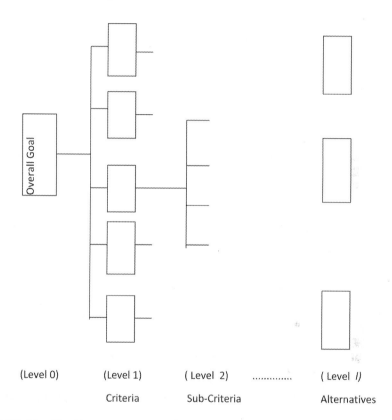

(Level 0)	(Level 1)	(Level 2)	....	(Level l)
	Criteria	Sub-Criteria		Alternatives

FIGURE 6.3   The hierarchical structure representation of a decision problem.

Step 2. DM provides pair-wise preferences at each level.

After the problem has been structured in the form of a hierarchy, the next step is to seek pair-wise preference judgments concerning the alternatives for the next higher-level sub-criteria (or criteria). These preference judgments are usually obtained from the DM in the form of a pair-wise comparison. The pair-wise comparisons can be provided using any appropriate ratio scale; a commonly used scale is the one proposed by Saaty as shown in Table 6.2. Where measured outcome achievements are available for any criterion or sub-criterion, then ratios of those measured values should be used instead of pair-wise comparisons from the DM.

Step 3. Determine the rank order of alternatives by computing at Level 0 the composite priority values of the alternatives.

The pair-wise preference judgments of alternatives or sub-criteria are used to compute the composite priorities at the next upper level. This process of computing composite priorities at the next upper level is continued till the composite priority values are obtained for Level 0. At each level, the composite values are the weighted average of composite values at the lower level of the hierarchy. The composite values at Level 0 can be used to rank order the alternatives.

Since there can be inconsistency in pair-wise judgments provided by the DM, the AHP method provides a measure of such inconsistency. We will use the Golden and Wang proposed measure of inconsistency, G, instead of the one proposed by Saaty because G is easier to compute. The AHP steps are explained below using Problem 6.2 data.

6.4.3.1   Solution of Problem 6.2 Using AHP Method

Step 1. Use any of the methods described in Chapter 3 to identify alternatives and criteria. Next, create the hierarchy structure.

The hierarchical structure for Problem 6.2 is shown in Figure 6.1.

Step 2. Pair-wise judgments, v_{ij}, are provided by DM for each level of hierarchy.

In the pair-wise comparison matrix, the number of rows always equals the number of columns. All the pair-wise comparisons for Problem 6.2 are shown in Tables 6.21, 6.22, and 6.23. The last column in each of these tables shows the computational results of Step 3. Note, w.r.t. means 'with respect to'.

TABLE 6.21
Level 1 Pair-wise Comparisons w.r.t. Goal (Level 0) of Problem 6.2

	C_1	C_2	C_3	g'	g_{10}
C_1	1	1/3	5	$(1*1/3*5)^{1/3} = 1.186$	$1.186/4.25 = 0.28$
C_2	3	1	7	$(3*1*7)^{1/3} = 2.759$	$2.759/4.25 = 0.65$
C_3	1/5	1/7	1	0.306	$0.306/4.25 = 0.07$
Σ	4.20	1.48	13.00	$1.186 + 2.759 + 0.362 = 4.25$	1

$G_{10} = 0.153$, $n_{10} = 3$, $N_s = 0.1204$. Inconsistency too high as $0.153 > 0.1204$

TABLE 6.22
Level 2 Pair-wise Comparisons w.r.t. C_2 of Problem 6.2

	D_1	D_2	g_{2C2}
D_1	1	2	0.667
D_2	1/2	1	0.337

$G_{2C2} = 0$, consistent (as it is perfect matrix)

TABLE 6.23
Level 3 Pair-wise Comparisons of Problem 6.2

With Respect to C_1

	A_1	A_2	A_3	g_{1C1}
A_1	1	2	1	0.413
A_2	1/2	1	1	0.260
A_3	1	1	1	0.327

$G_{1C1} = 0.159 > 0.1204$, too high inconsistency.
$N_{1C1} = 3$, $N_s = 0.1204$

With Respect to D_1

	A_1	A_2	A_3	g_{2D1}
A_1	1	2	3	0.540
A_2	1/2	1	2	0.297
A_3	1/3	1/2	1	0.163

$G_{2D1} = 0.064 < 0.1204$, acceptable inconsistency.
$N_{2D1} = 3$, $N_s = 0.1204$

With Respect to D_2

	A_1	A_2	A_3	g_{2D2}
A_1	1	2	1/7	0.131
A_2	1/2	1	1/9	0.076
A_3	7	9	1	0.793

$G_{2D2} = 0.061 < 0.1204$, acceptable inconsistency.
$N_{2D2} = 3$, $N_s = 0.1204$

With Respect to C_3

	A_1	A_2	A_3	g_{1C3}
A_1	1	2	4	0.558
A_2	1/2	1	3	0.320
A_3	1/4	1/3	1	0.122

$G_{1C3} = 0.089 < 0.1204$, acceptable inconsistency.
$N_{1C3} = 3$, $N_s = 0.1204$

*Step 3. Use the method proposed by Golden and Wang to compute **g** vectors and the corresponding inconsistency measures G.*

Tables 6.21, 6.22, and 6.23 show the computed **g** and G values. If any G value for any pair-wise comparison matrix indicates unacceptable inconsistency (i.e., $G > N_s$), we should go back to Step 2 for those judgments and request DM to provide revised judgments till consistency is achieved. The N_s values are obtained from Table 6.24, where *n* is the number of rows or columns in the comparison matrix of step 2.

TABLE 6.24
Maximum Acceptable Value of G for It to Be Consistent

n	3	4	5	6	7	8	9	10	11
N_s	0.1204	0.2032	0.2586	0.2991	0.3218	0.3442	0.3596	0.3729	0.3833

Step 3 computations are explained here (where columns are enumerated as j and rows are enumerated as i):

a. Compute Normalized Geometric Mean Vector, g, for each comparison matrix as shown below:

$$g' = \begin{bmatrix} g'_1 \\ g'_2 \\ \dots \\ g'_n \end{bmatrix} = \begin{bmatrix} \sqrt[n]{v_{11}*v_{12}\dots *v_{1n}} \\ \sqrt[n]{v_{21}*v_{22}\dots *v_{2n}} \\ \dots \\ \sqrt[n]{v_{n1}*v_{n2}\dots *v_{nn}} \end{bmatrix}$$

$$g = \begin{bmatrix} g_1 \\ g_2 \\ \dots \\ g_n \end{bmatrix} = \begin{bmatrix} \dfrac{g'_1}{\sum_{i=1}^{i=n} g'_i} \\ \dfrac{g'_2}{\sum_{i=1}^{i=n} g'_i} \\ \dots \\ \dfrac{g'_n}{\sum_{i=1}^{i=n} g'_i} \end{bmatrix}$$

See Table 6.21 for computational details of g' and g for Level 1of Problem 6.2.

b. Normalize the Comparison Matrix, V, to matrix R. A simple normalization scheme is to divide each v_{ij} by the sum of the corresponding column values, $\sum_{i=1}^{i=n} v_{ij}$.

See Table 6.25 for computational details of matrix R for Level 1 of Problem 6.2. Note, the values 4.20, 1.48, and 13.00 are from the last row of Table 6.21.

TABLE 6.25
Level 1 Normalized Comparison Matrix, R, of Problem 6.2

	R_1	R_2	R_3
R_1	1/4.20 = 0.238	0.226	5/13.00 = 0.385
R_2	3/4.20 = 0.714	1/1.48 = 0.677	0.538
R_3	0.20/4.20 = 0.048	0.097	0.077

c. Compute G as follows (where $|X|$ means the absolute value of X):

$$G = \frac{1}{n} * \left[\sum_{i=1}^{i=n} \left(\sum_{j=1}^{j=k} |r_{ij} - g_i| \right) \right]$$

Inconsistency is too high or unacceptable if G is greater than the N_s values shown in Table 6.24, where n is the number of rows (equal to the number of columns) of the normalized comparison matrix:

To compute the G of the matrix in Table 6.21, we first compute the normalized comparison matrix R as shown in Table 6.25. Also, from Table 6.24, for $n = 3$, $N_s = 0.1204$. Next, we compute G for Level 1 matrix,

$$G_{10} = \frac{1}{3} * [(|r_{11} - g_1| + |r_{12} - g_1| + |r_{13} - g_1|) + (|r_{21} - g_2| + |r_{22} - g_2| + |r_{23} - g_2|)$$
$$+ (|r_{31} - g_3| + |r_{32} - g_3| + |r_{33} - g_3|)] = \frac{1}{3} * [(|0.238 - 0.28| + |0.226 - 0.28|$$
$$+ |0.385 - 0.28|) + (|0.714 - 0.65| + |0.677 - 0.65| + |0.538 - 0.65|) + (|0.048$$
$$- 0.07| + |0.097 - 0.07| + |0.077 - 0.07|)] = \frac{1}{3} * [(0.042 + 0.054 + 0.105)$$
$$+ (0.064 + 0.027 + 0.112) + (0.022 + 0.027 + 0.007)] = 0.159.$$

But, as 0.159 is greater than 0.1204, it means inconsistency is too high. Therefore, we should go back to the DM to request a new set of pair-wise comparisons for Level 1 till the level of inconsistency is within an acceptable range. However, to keep the displayed number of computations manageable for this computational demonstration, we will skip that re-evaluation phase and keep this inconsistent result.

For level 2, Table 6.22 shows the computed $\mathbf{g_{2C2}}$ and G_{2C2} values for C_2.

For Level 3, Table 6.23 shows the computed $\mathbf{g_{1C1}}$ and G_{1C1} values for C_1, and $\mathbf{g_{1C3}}$ and G_{1C3} values for C_3.

Due to the complexity of the hierarchy of Problem 6.2 we must have additional computations for the $\mathbf{g_{1C2}}$ and G_{1C2} values for C_2. We proceed as follows:

i. from Table 6.23 the g and G values at D_1 and D_2 of the three alternatives at Level 2 are:

$\mathbf{g_{2D1}} = (0.540, 0.297, 0.163)$ and $G_{2D1} = 0.064 < 0.1204$, acceptable inconsistency; $n = 3$, $N_s = 0.1204$.

$\mathbf{g_{2D2}} = (0.131, 0.076, 0.793)$ and $G_{2D2} = 0.061 < 0.1204$, acceptable inconsistency; $n = 3$, $N_s = 0.1204$.

ii. we now compute the $\mathbf{g_{1C2}}$ and G_{1C2} values for C_2 at Level 1 (rolled up from Level 2) as follows:

$\mathbf{g_{1C2}}$(for A_1) $= \mathbf{g_{2C2}}$(for D_1)*$\mathbf{g_{2D1}}$(for A_1) $+ \mathbf{g_{2C2}}$(for D_2)*$\mathbf{g_{2D2}}$(for A_1) $=$ 0.667*0.540 + 0.337*0.131 = 0.404

$\mathbf{g_{1C2}}$(for A_2) = 0.667*0.297 + 0.337*0.076 = 0.224

$\mathbf{g_{1C2}}$(for A_3) = 0.667*0.163 + 0.337*0.793 = 0.376

We compute G for the above roll-up as follows:

$G_{1C2} = (\mathbf{g_{2C2}}*G_{2D1} + \mathbf{g_{2C2}}*G_{2D2}) = (0.667*0.064 + 0.337*0.061) = 0.063 <$ 0.1204, $n_{1C2} = 3$

Note that in Tables 6.21 and 6.23, wherever there is too high inconsistency in any set of pair-wise comparisons then we should ask the DM to revise one or more of those inconsistent pair-wise comparisons and then recalculate G, but, as we have indicated before, to reduce the recalculations in our demonstration example, we will skip the revision of comparisons. Computational results at Level 3 for C_1 and C_3 as shown in Table 6.23 are reproduced here:

$\mathbf{g_{1C1}}$ $(A_1, A_2, A_3) = (0.413, 0.260, 0.327)$, $G_{1C1} = 0.159 > 0.1204$, non-acceptable inconsistency; $n_{1C1} = 3$

$\mathbf{g_{1C3}}$ $(A_1, A_2, A_3) = (0.558, 0.320, 0.122)$, $G_{1C3} = 0.089 < 0.1204$, acceptable inconsistency; $n_{1C3} = 3$

iii. computational results for Level 1 of $\mathbf{g_{10}}$ and G_{10} as shown in Table 6.21 are reproduced below:

$\mathbf{g_{10}} = (0.28, 0.65, 0.07)$ and $G_{10} = 0.153 > 0.1204$, non-acceptable inconsistency; $n_0 = 3$

iv. after g values for different hierarchies have been computed and all G values have acceptable inconsistencies, we then compute the overall (or, composite) relative values of the alternatives $V(A_1)$, $V(A_2)$, and $V(A_3)$ as well as the corresponding inconsistency index G at Level 0 as follows:

$V(A_1) = \mathbf{g_{10}}(\text{for } C_1)*\mathbf{g_{1C1}}(\text{for } A_1) + \mathbf{g_{10}}(\text{for } C_2)*\mathbf{g_{1C2}}(\text{for } A_1) + \mathbf{g_{10}}(C_3)*\mathbf{g_{1C3}}(A_1)$
$\qquad = 0.28*0.413 + 0.65*0.404 + 0.07*0.558 = 0.418$

$V(A_2) = 0.28*0.260 + 0.65*0.224 + 0.07*0.320 = 0.241$

$V(A_3) = 0.28*0.327 + 0.65*0.376 + 0.07*0.122 = 0.344$

$$G_0 = \frac{\left(1*G_{10}\right) + \left(\mathbf{g_{10c_1}}*G_{1c_1} + \mathbf{g_{10c_2}}*G_{1c_2} + \mathbf{g_{10c_3}}*G_{1c_3}\right)}{\left(1 + \mathbf{g_{10c_1}} + \mathbf{g_{10c_2}} + \mathbf{g_{10c_3}}\right)}$$

$$= \frac{\left(1*0.153\right) + \left(0.280*0.159 + 0.65*0.063 + 0.07*0.089\right)}{\left(1 + 0.28 + 0.65 + 0.07\right)} = 0.1222$$

The dimension, n, of G_0 is computed as follows:

$$n = \frac{\left(1*n_{10} + \mathbf{g_{0c1}}*n_{1c1} + \mathbf{g_{0c2}}*n_{1c2} + \mathbf{g_{0c3}}*n_{1c3}\right)}{\left(1 + \mathbf{g_{0c1}} + \mathbf{g_{0c2}} + \mathbf{g_{0c3}}\right)}$$

$$= \frac{\left(1*3 + 0.28*3 + 0.65*3 + 0.07*3\right)}{\left(1 + 0.28 + 0.65 + 30.07\right)} = 3$$

$N_s = 0.1204$. $G > N_s$ indicates marginally high inconsistency. In other words, the high inconsistency we determined at a lower level (indicated in Tables 6.21 and 6.23) has percolated up to here.

We can now preference rank order the alternatives (based on their V) as A_1, A_3, A_2.

6.4.3.2 Solution of Problem 6.3 Using AHP Method

Before we solve Problem 6.3, we will review a few features of the problem.

a. The hierarchy of Problem 6.3 is simpler compared to that of Problem 6.2.
b. The outcome achievements of all five alternatives with respect to each criterion are based on the measured values using the scales of each criterion.

Therefore, the DM need not provide pair-wise comparisons of the alternatives in Level 2 with respect to each criterion as we can use the ratios of the measured values to determine the pair-wise comparisons. DM needs to provide the pair-wise comparisons of the three criteria in Level 1 only.

c. Since AHP ratios must be positive and greater than zero, we will normalize all the Table 6.1 values for the measurement scale ranges of each criterion using linear normalization. The measurement scale ranges are:

$$C_1: \left(\text{Highest, Lowest}\right) = (100,10), \; C_2: (10,-10), \; C_3: (10,0)$$

Table 6.26 shows the (Linear) normalized values.

TABLE 6.26

Normalized Pay-off Matrix for Problem 6.3

	R_1	R_2	R_3
A_1	$(60-10)/(100-10) = 0.55556$	$(-3+10)/10+10) = 0.35000$	0.70000
A_2	0.31111	0.90000	0.50000
A_3	0.80000	0.25000	1.00000
A_4	0.51111	0.60000	0.60000
A_5	0.21111	0.50000	0.90000

Step 1. The hierarchical structure for Problem 6.3 is shown in Figure 6.2.

Step 2. Table 6.27 shows the pair-wise judgments, v_{ij}, provided by DM for Level 1 of the hierarchy. Table 6.27 also shows computed $\mathbf{g_{10}}$ and G_{10} at Level 1.

TABLE 6.27

Level 1 Computation of g_{10} and G_{10} for Problem 6.3

	C_1	C_2	C_3	g'	g_{10}
C_1	1	1/3	5	1.1856	1.1856/4.307 = 0.28
C_2	3	1	7	$(3*1*7)^{1/3} = 2.7589$	2.7589/4.307 = 0.64
C_3	1/3	1/7	1	0.3624	0.08
Σ	4.33	1.48	13	4.307	

	R_1	R_2	R_3
C_1	1/4.33 = 0.23	0.33/1.48 = 0.23	5/13 = 0.38
C_2	0.69	0.68	0.54
C_3	0.08	0.10	0.08

$G_{10} = 0.04$, $N_s = 0.1204$, n = 3. Consistent

G_{10} = ABS((0.23 – 0.28) + (0.23 – 0.28) + (0.38 – 0.28) + (0.69 – 0.64) + (0.68 – 0.64) + (0.54 – 0.64) + (0.08 – 0.08) + (0.10 – 0.08) + (0.08 – 0.08))/3 = 0.04

The pair-wise comparisons for Level 2 in Table 6.28 are computed using the Table 6.26 normalized pay-off matrix values.

TABLE 6.28
Level 2 Pair-wise Comparisons for Problem 6.3

| | With Respect to C_1 | | | | | |
	A_1	A_2	A_3	A_4	A_5	g_{2c1}
A_1	1.00	0.55556/ 0.31111 = 1.79	0.55556/ 0.80000 = 0.69	1.09	0.55556/ 0.21111 = 2.63	0.23
A_2	0.31111/ 0.55556 = 0.56	1.00	0.39	0.61	1.47	0.13
A_3	1.44	2.57	1.00	1.57	3.79	0.33
A_4	0.92	1.64	0.64	1.00	2.42	0.21
A_5	0.38	0.68	0.26	0.41	1.00	0.11

G_{2c1} = 0.0, N_s = 0.2586, n = 5. Consistent

| | With Respect to C_2 | | | | | |
	A_1	A_2	A_3	A_4	A_5	g_{2c2}
A_1	1.00	0.39	1.40	0.58	0.70	0.13
A_2	2.57	1.00	3.60	1.50	1.80	0.35
A_3	0.71	0.28	1.00	0.42	0.50	0.10
A_4	1.71	0.67	2.40	1.00	1.20	0.23
A_5	1.43	0.56	2.00	0.83	1.00	0.19

G_{2c2} = 0, N_s = 0.2586, n = 5. Consistent

| | With Respect to C_3 | | | | | |
	A_1	A_2	A_3	A_4	A_5	g_{2c3}
A_1	1.00	1.40	0.70	1.17	0.78	0.19
A_2	0.71	1.00	0.50	0.83	0.56	0.14
A_3	1.43	2.00	1.00	1.67	1.11	0.27
A_4	0.86	1.20	0.60	1.00	0.67	0.16
A_5	1.29	1.80	0.90	1.50	1.00	0.24

G_{2c3} = 0, N_s = 0.2586, n = 5. Consistent

Step 3. The last column in Tables 6.27 and 6.28 shows the computed *g*; *G* values are shown in the last rows. Note that all table values shown are rounded (up or down) to two decimal digits. In Table 6.28, all the computed *G* values are equal to zero. The reason is in how the pair-wise comparisons have been determined as opposed to those in Table 6.27 that have been obtained from the DM, with potential for inconsistency in DM's responses. The Table 6.28 values have been computed by using the measured values in the normalized pay-off matrix in Table 6.26. An example of this computation is also shown in the table. Since actual measured values are used for computing these ratios, there is no scope for inconsistency and, hence, *G* = 0.

The computed overall (or, composite) relative values of the alternatives $V(A_1)$, $V(A_2)$, $V(A_3)$, $V(A_4)$, and $V(A_5)$ as well as the corresponding inconsistency index G at Level 0 are shown below:

$V(A_1) = g_0(C_1)*g_{C1}(A_1) + g_0(C_2)*g_{C2}(A_1) + g_0(C_3)*g_{C3}(A_1) = 0.28*0.23 + 0.64*0.13 + 0.08*0.19 = 0.164$

$V(A_2) = 0.268$

$V(A_3) = 0.174$

$V(A_4) = 0.219$

$V(A_5) = 0.175$

We have not computed G here because it will be very close to 0 given G values in Tables 6.27 and 6.28 are either 0 or very close to 0.

The rank order of the alternatives is = A_2, A_4, (A_3 and A_5), A_1.

Postscript

This book has been written as a guide for all managers at different levels of any organization to improve their decision-making ability based on a quantitative approach. The process of decision-making has been presented from a holistic point of view where not only have we provided some easy to apply decision methods but also the tools for problem analysis, for development of decision alternatives and comparison criteria, and measurement of criteria achievement. The tools and techniques that we have provided are not state-of-the-art research results but are the fundamental ones that can be considered as tried and true. Our selection of the tools and techniques is based on those that we have found to be easily understood and applied by most of our seminar and workshop participants with various levels of mathematical background and computer literacy from different types of organizations in the US and internationally. None of the tools and techniques that we have explained is based on our original research. They have been developed and proposed by various researchers and professionals. The references listed in this book are the primary source for these materials. Many websites and digital sources also describe and explain some of these tools and techniques as they relate to various application disciplines. Everything that we have presented can very easily be applied using any computer-based spreadsheet program. A simple Internet search will also reveal the availability of many free and fee-based software specific to some of the tools. Our sincere hope is that after reading this book a decision-maker would be able to pick any tool described in the book and then be able to apply that in their managerial decision situations. We sincerely hope that you will be able to reach a better decision by using an appropriate tool or technique described in this book.

References

Bunn, D. W. 1984. *Applied Decision Analysis*, New York, NY: McGraw-Hill.

Golden, B. and Q. Wang. 1989. An Alternate Measure of Consistency. In *The Analytic Hierarchy Process: Applications and Studies*, eds. B. Golden, E. Wasil, and P. Harker, 66–81. New York, NY: Springer-Verlag.

Hwang, C. L. and A. Masud. 1979. *Multiple Objective Decision Making—Methods and Applications*. New York, NY: Springer-Verlag.

Hwang, C. L. and K. S. Yoon. 1981. *Multiple Attribute Decision Making—Methods and Applications*. New York, NY: Springer-Verlag.

Ishikawa, K. 1982. *Guide to Quality Control*, New York, NY: Quality Resources.

Masud, A. and A. Ravindran, 2008. Multiple Criteria Decision Making. In *Operations Research and Management Science Handbook*, ed. A. Ravindran, 5.1–5.41. New York, NY: CRC Press.

Morris, W. T. 1977. *Decision Analysis*, Columbus, OH: Grid Inc.

Saaty, T. 1990. *Multicriteria Decision Making: The Analytic Hierarchy Process*, Pittsburgh, PA: RWS Publications.

Index

Printed in the United States
by Baker & Taylor Publisher Services